ENVIRONMENTALSTATS for S-PLUS®

Second Edition

Springer
*New York
Berlin
Heidelberg
Barcelona
Hong Kong
London
Milan
Paris
Singapore
Tokyo*

Steven P. Millard

ENVIRONMENTALSTATS for S-PLUS®

User's Manual for Version 2.0

Second Edition

Springer

Steven P. Millard
Probability, Statistics, and Information
7723 44th Avenue NE
Seattle, WA 98115-5117
USA
smillard@probstatinfo.com

Library of Congress Cataloging-in-Publication Data
Millard, Steven P.
    EnvironmentalStats for S-Plus : user's manual for Version 2.0 / Steven P. Millard—
2nd ed.
        p. cm.
    Includes bibliographical references and index.
    ISBN 0-387-95398-1 (s/c : alk. paper)
    1. Environmental sciences—Statistical methods—Data processing. 2. S-Plus. I. Title
GE45.S73 M52   2002
363.7'007'27—dc21   2001058264

Printed on acid-free paper.

S-PLUS is a registered trademark of Insightful Corporation.

Production managed by Steven Pisano; manufacturing supervised by Jerome Basma.
Photocomposed pages prepared from the author's Microsoft Word files.
Printed and bound by Maple-Vail Book Manufacturing Group, York, PA.
Printed in the United States of America.

9 8 7 6 5 4 3 2 1

ISBN 0-387-95398-1   SPIN 10858683

Springer-Verlag   New York Berlin Heidelberg
A member of BertelsmannSpringer Science+Business Media GmbH

Preface

The environmental movement of the 1960s and 1970s resulted in the creation of several laws aimed at protecting the environment, and in the creation of Federal, state, and local government agencies charged with enforcing these laws. Most of these laws mandate monitoring or assessment of the physical environment, which means someone has to collect, analyze, and explain environmental data. Numerous excellent journal articles, guidance documents, and books have been published to explain various aspects of applying statistical methods to environmental data analysis. Also, several specialty software packages for specific niches in environmental statistics exist. Not very many software packages provide a comprehensive treatment of environmental statistics in general.

ENVIRONMENTALSTATS for S-PLUS is a new *comprehensive* software package for environmental statistics. An add-on module to the statistical software package S-PLUS (from Insightful Corporation), it provides a set of powerful yet simple-to-use menu items and functions for performing graphical and statistical analyses of environmental data. ENVIRONMENTALSTATS for S-PLUS brings the major environmental statistical methods found in the literature and regulatory guidance documents into one statistical package, along with an extensive hypertext help system that explains *what* these methods do, *how* to use these methods, and *where* to find them in the environmental statistics literature. Also included are numerous built-in data sets from regulatory guidance documents and the environmental statistics literature. ENVIRONMENTALSTATS for S-PLUS, combined with S+SPATIALSTATS and S-PLUS for ArcView GIS, provide the environmental scientist, statistician, researcher, and technician with all the tools to "get the job done!"

ENVIRONMENTALSTATS for S-PLUS and this User's Manual are intended for anyone who has to make sense of environmental data, including statisticians, hydrologists, soil scientists, atmospheric scientists, geochemists, environmental engineers and consultants, hazardous and solid waste site managers, and regulatory agency analysts and enforcement officers. Some parts of ENVIRONMENTALSTATS for S-PLUS incorporate statistical methods that have appeared in the environmental literature but are not commonly found in any statistical software package. Some parts are specifically aimed at users who are required to collect and analyze environmental monitoring data in order to comply with federal and state Superfund, RCRA, CERCLA, and Subtitle D regulations for environmental monitoring at hazardous and solid waste sites. All of the functions in ENVIRONMENTALSTATS for S-PLUS, however, are useful to anyone who needs to analyze environmental data.

This manual is divided into 9 chapters. *Chapter 1: Getting Started* is an introduction to environmental statistics in general and ENVIRONMENTALSTATS for S-PLUS in particular, and includes information on system and user requirements, installing the software, loading and using the module, and getting technical support. The last section of the chapter includes a tutorial.

Chapters 2 – 8 contain information about how to use the menus and functions in ENVIRONMENTALSTATS for S-PLUS to design sampling programs and perform graphical and statistical analyses of environmental data. Finally, Chapter 9 shows you how to use ENVIRONMENTALSTATS for S-PLUS to perform Monte Carlo simulation and probabilistic risk assessment.

At the back of the book is an extensive list of references for environmental statistics as well as index of key words and terms. In addition to using the index, you are encouraged to use the on-line hypertext help system as well.

Typographic Conventions

Throughout this User's Manual, the following typographic conventions are used:

- Menu selections are shown in an abbreviated form using the **bold font** and the arrow symbol (>) to indicate a selection within a menu, as in **File>New**.
- The **bold font** is also used to display what you type within menu text boxes, what menu button to choose, what menu check box to check, etc., and also for Windows commands and filenames, as well as for chapter and section headings. Sometimes it is also used for emphasis.
- The `courier font` is used to display the names of S-PLUS objects, what you type within an S-PLUS Command or Script Window, and output from S-PLUS.
- The *italic font* is used for chapter and help file titles, emphasis, and user-supplied variables within S-PLUS commands.
- The ***bold italic font*** is used for emphasis.

Within chapters, S-PLUS commands are preceded with the "greater than" sign, i.e., >, which is the default S-PLUS prompt. For commands that require more than one line of input, the line or lines following the first line are indented, whereas within S-PLUS they are preceded with the "plus" sign, i.e., +, which is the default S-PLUS continuation prompt. Note that page size and formatting for this book determine how command lines are split. As a user, you can choose to split lines differently or not at all.

A Note About Graphics

All of the graphs you can create with ENVIRONMENTALSTATS for S-PLUS use traditional S-PLUS graphics; thus they are editable from the Command or Script Window but not editable using the Graphical User Interface (GUI). If you use the Windows version of S-PLUS, you can convert traditional plots to GUI-editable plots by right-clicking on the data part of the graph and choosing **Convert to Objects** from the context menu. In all of the examples that produce plots with legends, if you want to produce the plot using the ENVIRONMENTALSTATS for S-PLUS Menu (rather than the Command or Script Window), then after following the instructions for producing the plot with the Menu, you will have to convert the plot to a GUI-editable plot, then insert the legend. (You will also have to follow this procedure if you use the ENVIRONMENTALSTATS for S-PLUS Menu to create a plot without a title and then want to add one.)

It is very easy to produce color plots in S-PLUS and ENVIRONMENTALSTATS for S-PLUS. Many of the built-in plotting functions produce color plots by default. In this User's Manual, however, all of the plots are black-and-white or gray-scale due to the high cost of color printing. Instructions for producing color plots are still included in the examples, but the pictures in this User's Manual will be in black-and-white, whereas the pictures on your computer screen will be in color.

Reproducing Examples in the Help Files

In the ENVIRONMENTALSTATS for S-PLUS help files, S-PLUS commands are shown with the default S-PLUS prompt (>) and continuation prompt (+), so if you copy S-PLUS commands displayed in a help file and paste them into the S-PLUS Command Window, S-PLUS will indicate syntax errors. You can use the Command History tool under the Tools menu to select the lines just pasted into S-PLUS, edit them to remove the S-PLUS prompts, and re-execute them. Alternatively, you can instead paste the lines into a Script Window and edit the lines before executing them from the Script Window.

Companion Textbook

This User's Manual provides brief explanations of various topics in environmental statistics and explains how to use ENVIRONMENTALSTATS for S-PLUS. A companion textbook, **Environmental Statistics with S-PLUS** by Steven P. Millard and Nagaraj K. Neerchal (CRC Press, 2001), provides more details about various topics in environmental statistics and can be used as a textbook for a course in environmental statistics.

Technical Support

For technical support and information on S-PLUS, ENVIRONMENTALSTATS for S-PLUS, and other S-PLUS modules, please contact Insightful Corporation:

Insightful Corporation
1700 Westlake Ave N, Suite 500
Seattle, WA 98109-3044 USA
800-569-0123
support@insightful.com
www.insightful.com

Insightful Corporation
Knightway House
Park Street
Bagshot, Surrey
GU19 5AQ
United Kingdom
+44 1276 452 299
support@uk.insightful.com
www.uk.insightful.com

Acknowledgments

In the early 1980s, while pursuing a degree in biostatistics, I became aware of a knowledge gap in the field of environmental statistics. There was lots of research going on in the academic field of environmental statistics, but there were lots of poor designs and analyses being carried out in the real-world "field." There were even federal laws mandating incorrect statistical analyses of ground water samples at hazardous waste sites.

One of the first steps to improving the quality of environmental statistics is to improve the quality of communication between statisticians and experts in various environmental fields. Several researchers and practitioners have done this by publishing excellent books and journal articles dealing with general and specific problems in environmental statistics (see the ENVIRONMENTALSTATS for S-PLUS help topic *References for Environmental Statistics*).

The next logical step is to provide the necessary tools to carry out all the great methods and ideas in the literature. In the mid-1980s I sat down with two friends of mine, Dennis Lettenmaier and Jim Hughes, to talk about building a software package for environmental statistics. Dennis and Jim gave me some good ideas, but I never acted on these ideas because I didn't feel like I had the tools to make the kind of software package I wanted to.

The emergence of S-PLUS and help-authoring tools changed all that. S-PLUS, from Insightful Corporation, is a premiere statistical software package with great graphics that allows users to write their own functions and create pull-down menus. ForeHelp, from ForeFront, Inc., is one of several help-authoring tools that allows anyone who can figure out a word processor to write a hypertext Windows, HTML, or JAVA help system. The availability of these tools resulted in the creation of ENVIRONMENTALSTATS for S-PLUS.

There are several people who have helped and encouraged me over the past several years as I have developed and extended ENVIRONMENTALSTATS for S-PLUS. First of all, thanks to Insightful for marketing, distributing, and supporting this module! Several people at Insightful have contributed to the success of ENVIRONMENTALSTATS for S-PLUS, including Kim Leader and Cheryl Mauer assisting me with development schedule, Patrick Aboyoun assisting me with the Graphical User Interface, Tim Wegner assisting me with the help system, and Rich Calaway, Charlie Roosen, Stephen Kaluzny, and the Insightful technical support team answering my numerous questions. Special thanks to the members of the S-news group for constantly giving me insight into the workings of S-PLUS.

I would also like to thank Gilbert FitzGerald, former Director of Product Development at MathSoft for his encouragement and advice. Thanks to Jim Hughes (University of Washington) for being a statistical consultant's statistical consultant. Also thanks to Tim Cohn and Dennis Helsel (US Geological Survey), Chris Fraley (Insightful and University of Washington), Jon Hosking (IBM Research Division), Ross Prentice (Fred Hutchinson Cancer Research Center), and Terry Therneau (Mayo Clinic) for their help in answering technical questions. Dick Gilbert has been a constant source of encouragement and advice; thanks! Thanks to the Beta testers for their feedback on early versions.

There would be no ENVIRONMENTALSTATS for S-PLUS if there was no S-PLUS. There would be no S-PLUS if there had not first been S. I am grateful to the researchers of the Bell Laboratories S team at AT&T (now Lucent Technologies) who first created S, including Richard A. Becker, John M. Chambers, Alan R. Wilks, William S. Cleveland, and others. I am also grateful to Doug Martin for leading the creation of S-PLUS, and to Insightful for constantly enhancing and improving S-PLUS.

I am grateful to John Kimmel and Steve Pisano at Springer-Verlag for their help in transforming this User's Manual into a book. Finally, and most gratefully, thanks to my wife Stacy and my son Chris for their moral and financial support of this project. I couldn't have done it without you!

ENVIRONMENTALSTATS for S-PLUS is the culmination of a dream I had over 15 years ago. I hope it provides you with the tools you need to "get the job done," and I hope you enjoy it as much as I have enjoyed creating it!

Steven P. Millard
Seattle, October 2001

Contents

1

Getting Started

1.1 Introduction

Welcome to ENVIRONMENTALSTATS for S-PLUS! This User's Manual provides step-by-step guidance to using this software. ENVIRONMENTALSTATS for S-PLUS is an S-PLUS module for environmental statistics. This chapter is an introduction to environmental statistics in general and ENVIRONMENTALSTATS for S-PLUS in particular, and includes information on system and user requirements, installing the software, loading and using the module, and getting technical support. The last section of the chapter is a tutorial.

1.2 What is Environmental Statistics?

Environmental statistics is simply the application of statistical methods to problems concerning the environment. Examples of activities that require the use of environmental statistics include:

- Monitoring air or water quality.
- Monitoring groundwater quality near a hazardous or solid waste site.
- Using risk assessment to determine whether a potentially contaminated area needs to be cleaned up, and, if so, how much.
- Assessing whether a previously contaminated area has been cleaned up according to some specified criterion.
- Using hydrological data to predict the occurrences of floods.

The term "environmental statistics" must also include work done in various branches of ecology, such as animal population dynamics and general ecological modeling, as well as other fields, such as geology, chemistry, epidemiology, oceanography, and atmospheric modeling. This User's Manual concentrates on statistical methods to analyze chemical concentrations and physical parameters, usually in the context of mandated environmental monitoring.

Environmental statistics is a special field of statistics. Probability and statistics deal with situations in which the outcome is not certain. They are built upon the concepts of a *population* and a *sample* from the population. *Probability* deals with predicting the characteristics of the sample, given that you know the characteristics of the population (e.g., the probability of picking an ace out of a deck of 52 well-shuffled standard playing cards). *Statistics* deals with inferring the characteristics of the population, given information from one or more samples from the population (e.g., after 100 times of randomly choosing a card from a deck of 20 unknown playing cards and then replacing the card in the deck, no ace has appeared; therefore the deck probably does not contain any aces).

The field of environmental statistics is relatively young and employs several statistical methods that have been developed in other fields of statistics, such as sampling design, exploratory data analysis, basic estimation and hypothesis testing, quality control, multiple comparisons, survival analysis, and Monte Carlo simulation. Nonetheless, special problems have motivated innovative research, and both traditional and new journals now report on statistical methods that have been developed in the context of environmental monitoring. (See the help topic *References: Environmental Statistics—General References* for a list of general text books on environmental statistics.)

In addition, environmental legislation such as the Clean Water Act, the Clean Air Act, the Comprehensive Emergency Response, Compensation, and Liability Act (CERCLA), and the Resource and Recovery Act (RCRA) have spawned environmental regulations and agency guidance documents that mandate or suggest various statistical methods for environmental monitoring (see the help topic *References: Guidance Documents and Regulations*).

1.3   What is ENVIRONMENTALSTATS for S-PLUS?

ENVIRONMENTALSTATS for S-PLUS, created by Dr. Steven P. Millard of Probability, Statistics & Information (PSI), is a new *comprehensive* software package for environmental statistics. An add-on module to the statistical software package S-PLUS (from Insightful Corporation), it provides a set of powerful yet simple-to-use menu items and functions for performing graphical and statistical analyses of environmental data. ENVIRONMENTALSTATS for S-PLUS brings the major environmental statistical methods found in the literature and regulatory guidance documents into one statistical package, along with an extensive hypertext help system that explains *what* these methods do, *how* to use these methods, and *where* to find them in the environmental statistics literature. Also included are numerous built-in data sets from regulatory guidance documents and the environmental statistics literature. ENVIRONMENTALSTATS for S-PLUS, combined with S+SPATIALSTATS and S-PLUS for ArcView GIS, provide the environmental scientist, statistician, researcher, and technician with all the tools needed to "get the job done!"

Because ENVIRONMENTALSTATS for S-PLUS is an S-PLUS module, you automatically have access to all the features and functions of S-PLUS, including easy-to-use pull-down menus, powerful graphics, standard hypothesis tests, and the flexibility of a programming language. In addition, with ENVIRONMENTALSTATS for S-PLUS you can:

- Compute quantities associated with probability distributions (probability density functions, cumulative distribution functions, and quantiles), and generate random numbers from these distributions. (Several distributions have been added to the ones already available in S-PLUS.)
- Plot probability distributions so you can see how they change with the value of the distribution parameter(s).
- Compute several different kinds of summary statistics.
- Estimate distribution parameters and quantiles and compute confidence intervals for commonly used probability distributions.
- Perform and plot the results of goodness-of-fit tests.
- Compute optimal Box-Cox data transformations.
- Compute parametric and non-parametric prediction and tolerance intervals (including simultaneous prediction intervals).
- Perform additional hypothesis tests not already part of S-PLUS, including Chen's t-test for skewed distributions, Fisher's one-sample randomization test for location, the quantile test to detect a shift in the tail of one population relative to another, two-sample linear rank tests, the von Neumann rank test for serial correlation, and Kendall's seasonal test for trend.
- Perform power and sample size computations and create associated plots.
- Perform calibration based on a machine signal to determine decision and detection limits, and report estimated concentrations along with confidence intervals.
- Analyze singly and multiply censored (less-than-detection-limit) data with empirical cdf and Q-Q plots, parameter/quantile estimation and confidence intervals, prediction and tolerance intervals, goodness-of-fit tests, optimal Box-Cox transformations, and two-sample rank tests.
- Perform probabilistic risk assessment.
- Look up statistical methods in the environmental literature in a hypertext help system that explains the equations, links the equations to the original reference, includes abstracts of selected references, and contains a glossary of statistical and environmental terms.
- Reproduce specific examples in EPA guidance documents by using built-in data sets from these documents.

1.4   Intended Audience/Users

ENVIRONMENTALSTATS for S-PLUS and this User's Manual are intended for anyone who has to make sense of environmental data, including statisticians, environmental scientists, hydrologists, soil scientists, atmospheric scientists, geochemists, environmental engineers and consultants, hazardous and solid waste site managers, and regulatory agency analysts and enforcement officers. Some parts of ENVIRONMENTALSTATS for S-PLUS incorporate statistical methods that have appeared in the environmental literature but are not commonly found in any statistical software package. Some parts are specifically aimed at users who are required to collect and analyze environmental monitoring data in order to comply with federal and state Superfund, RCRA, CERCLA, and Subtitle D regulations for environmental monitoring at hazardous and solid waste sites. All of the functions in ENVIRONMENTALSTATS for S-PLUS, however, are useful to anyone who needs to analyze environmental data.

ENVIRONMENTALSTATS for S-PLUS is an S-PLUS module. In order to use it, you need to know how to perform basic operations in S-PLUS, such as using the pull-down menu or Command and Script Windows, reading data into S-PLUS, and creating basic data objects (e.g., data frames). See the S-PLUS documentation for more information on S-PLUS. In addition, you need to have a basic knowledge of probability and statistics. The User's Manual and the help system are not meant to be a text book in environmental statistics. See the references listed in *References: Environmental Statistics—General References* for a list of useful books and articles on this subject.

1.5   System Requirements

Because ENVIRONMENTALSTATS for S-PLUS is an S-PLUS module, it runs under every operating system that S-PLUS runs under, including Windows and UNIX. Currently only a Windows version is available. Windows users must have S-PLUS 6.0 in order to run EnvironmentalStats for S-PLUS Version 2.0. The module requires approximately 20 MB of space on your hard disk.

1.6   Installing ENVIRONMENTALSTATS for S-PLUS

Before you install EnvironmentalStats for S-PLUS, you should make sure you are using S-PLUS 6.0 or a later version. To determine what version of S-PLUS you are using, start S-PLUS and either choose **Help>About S-PLUS** on the menu, or type `version` at the S-PLUS prompt. If you are using the Professional version of S-PLUS for Windows, you should see something like the following:

```
> version
Professional Edition Version 6.0.2 Release 2 for
Microsoft Windows : 2001
```

To install ENVIRONMENTALSTATS for S-PLUS, close all Windows applications, and then simply run the Setup program included on the ENVIRONMENTALSTATS for S-PLUS distribution CD. To leave Setup before the setup is complete, select Cancel from any Setup dialog box. When you do this, Setup displays an exit dialog box that warns you that you have not completed the setup procedure.

1.7   Starting ENVIRONMENTALSTATS FOR S-PLUS

To start ENVIRONMENTALSTATS for S-PLUS, you must have already started S-PLUS. You can load the module either from the menu or from the Command Window.

- To load the module from the menu, on the S-PLUS menu bar make the following selections: **File>Load Module**. This brings up the Load Module dialog box. In the Module box, select **envstats**. For the Action buttons, select **Load Module**. Make sure the **Attach at top of search list** box is checked. Click **OK**.
- To load the module from the Command Window, type `module(envstats)` at the S-PLUS prompt.

Loading the module adds the EnvironmentalStats menu to the S-PLUS menu bar and also attaches the library of ENVIRONMENTALSTATS for S-PLUS functions to the second position in your search list (S-PLUS Professional users can type `search()` at the S-PLUS prompt to see the listing of directories on the search list). S-PLUS Standard and S-PLUS Professional users can access functions via the ENVIRONMENTALSTATS for S-PLUS pull-down menu. S-PLUS Professional users can access any of the functions in the ENVIRONMENTALSTATS for S-PLUS module via the command line as well.

Note: Some of the functions in ENVIRONMENTALSTATS for S-PLUS mask built-in S-PLUS functions. The masked functions are modified versions of the built-in functions and have been created to support the other functions in ENVIRONMENTALSTATS for S-PLUS, but the modifications should not affect normal use of S-PLUS. If you experience unexpected behavior of S-PLUS after attaching ENVIRONMENTALSTATS for S-PLUS, try unloading the module (see *Unloading ENVIRONMENTALSTATS for S-PLUS* below). All of the functions in ENVIRONMENTALSTATS for S-PLUS, whether they mask built-in functions or not, are described in the help system.

1.8   Getting Help

You may start the help system for ENVIRONMENTALSTATS for S-PLUS in one of five ways:

1.   On the menu bar, select **Help>Available Help>EnvironmentalStats**.
2.   On the menu bar, go to the **EnvironmentalStats** menu and select an item. A dialog box will be displayed. In the lower right-hand corner click on the **Help** button.
3.   At the command line, type `help(module="envstats")`.
4.   At the command line, use the ? operator and `help` function just as you do in S-PLUS. For example, type `?pdfplot` to call up the help file for the `pdfplot` function.

5.   Outside of S-PLUS, select **Start>Programs>S-PLUS 6.0 Professional>EnvironmentalStats**, or in the S-PLUS program group double-click the ENVIRONMENTALSTATS icon.

In the Table of Contents, you will see the main help categories, including Pull-Down Menu, Functions, Datasets, References, and Glossary. You can access the help files for functions by category or alphabetically, and you can access data sets by source or alphabetically.

The help system for ENVIRONMENTALSTATS for S-PLUS is a separate application from the help system for S-PLUS. Words that are underlined and highlighted in color (green or blue or whatever color your help system uses for jumps) are jumps to other help windows *within* the ENVIRONMENTALSTATS for S-PLUS help system. In each help file, if a word appears under the paragraph heading SEE ALSO but is not underlined and highlighted, you must use the S-PLUS help system to look up this word.

For example, in the help system for ENVIRONMENTALSTATS for S-PLUS, in the Table of Contents choose **Functions>Functions by Category**. This will bring up the help file *EnvironmentalStats for S-PLUS Functions By Category*. Click on *Plotting Probability Distributions* and then click on `pdfplot`. Within that help file, under the paragraph heading SEE ALSO, you will see:

```
.Distribution.frame, Probability Distributions and Random
    Numbers, cdfplot, ecdfplot, qqplot, qqplot.gestalt,
    plot, plot.default, par, title.
```

You can access the help files for `.Distribution.frame`, `Probability Distributions and Random Numbers`, `cdfplot`, `ecdfplot`, `qqplot`, and `qqplot.gestalt` directly by simply clicking on these words. To look at the help file for the functions `plot`, `plot.default`, `par`, and `title`, however, you must use the S-PLUS help system.

1.9   Customizing ENVIRONMENTALSTATS for S-PLUS

If you plan to use ENVIRONMENTALSTATS for S-PLUS extensively, you may want to customize your S-PLUS startup routine to automatically attach the ENVIRONMENTALSTATS for S-PLUS module each time you start S-PLUS. S-PLUS Professional users can do this by adding the line `module(envstats)` to your `.First` function. If you do not have a `.First` function, you can create one by simply typing the following command:

```
> .First <- function() {module(envstats)}
```

If you want to automatically attach the ENVIRONMENTALSTATS for S-PLUS module each time you start S-PLUS *and* start the ENVIRONMENTALSTATS for S-PLUS help system, make sure your `.First` function includes the following lines:

```
module(envstats)
help(module="envstats")
```

If you are going to use ENVIRONMENTALSTATS for S-PLUS for several different projects, it is a good idea to use separate directories for each project. See the S-PLUS documentation for how to create separate "Chapters" and have S-PLUS ask you which directory you want to work in when you start S-PLUS.

1.10 Unloading ENVIRONMENTALSTATS for S-PLUS

To remove the ENVIRONMENTALSTATS for S-PLUS module from your S-PLUS session, S-PLUS Professional users can simply type the following command at the S-PLUS prompt:

```
> module(envstats, unload=T)
```

This command removes the ENVIRONMENTALSTATS for S-PLUS menu from the S-PLUS menu bar and the library of ENVIRONMENTALSTATS for S-PLUS functions from your search list.

1.11 A Tutorial

This section is a brief tutorial that highlights some of the major features of ENVIRONMENTALSTATS for S-PLUS. There are several ways to use this section. If you are fairly new to S-PLUS, you may want to briefly skim this section to get an idea of what you can do in ENVIRONMENTALSTATS for S-PLUS, and then come back later after you have read the other chapters of this manual. If you have used S-PLUS for a long time and have just installed ENVIRONMENTALSTATS for S-PLUS, you may want to follow this tutorial in depth right now to get acquainted with some of the features available in this S-PLUS module. Throughout this section we assume you have started S-PLUS and also loaded ENVIRONMENTALSTATS for S-PLUS.

1.11.1 The TcCB Data

The guidance document *Statistical Methods for Evaluating the Attainment of Cleanup Standards, Volume 3: Reference-Based Standards for Soils and Solid Media* (USEPA, 1994b, pp.6.22-6.25) contains measures of 1,2,3,4-Tetrachlorobenzene (TcCB) concentrations (ppb) from soil samples at a Reference site and a Cleanup area. There are 47 observations from the Reference site and 77 in the Cleanup area. These data are stored in the data frame `epa.94b.tccb.df` (see the help file *Datasets: USEPA (1994b)*). There is one observation coded as "ND" in this data set as presented in the guidance document. Here, we'll assume this observation is less than the smallest observed value, which is 0.09 ppb. For the purposes of this tutorial, we'll set this one censored observation to the assumed detection limit of 0.09.

To look at the raw data, open the Object Explorer Window, click on the **Find Objects** button (binoculars icon), type **epa.94b.tccb.df** and click **OK**. In the Object Explorer Window, under the Object column, double-click on **epa.94b.tccb.df**. This will bring up a data sheet. Alternatively, in the Command Window, type epa.94b.tccb.df:

```
> epa.94b.tccb.df
      TcCB.orig   TcCB Censored      Area
    1      0.22   0.22        F Reference
    2      0.23   0.23        F Reference
    .
  124    168.64 168.64        F Cleanup
```

For the remainder of this tutorial, we will assume that you have attached the data frame epa.94b.tccb.df to your search list with the following command:

```
> attach(epa.94b.tccb.df)
```

Note that if you are only going to follow the instructions for using the Menu you do not have to execute the above command.

1.11.2 Computing Summary Statistics

The summary statistics for the TcCB data are shown below.

```
                                  Cleanup Reference
                    Sample Size:  77        47
                     # Missing:   0         0
                          Mean:   3.915     0.5985
                        Median:   0.43      0.54
              10% Trimmed Mean:   0.6846    0.5728
                Geometric Mean:   0.5784    0.5382
                          Skew:   7.717     0.9019
                      Kurtosis:   62.67     0.132
                           Min:   0.09      0.22
                           Max: 168.6       1.33
                         Range: 168.6       1.11
                  1st Quartile:   0.23      0.39
                  3rd Quartile:   1.1       0.75
            Standard Deviation:   20.02     0.2836
  Geometric Standard Deviation:   3.898     1.597
           Interquartile Range:   0.87      0.36
      Median Absolute Deviation:  0.3558    0.2669
      Coefficient of Variation:   5.112     0.4739
```

These summary statistics indicate that the observations for the Cleanup area are extremely skewed to the right. The medians for the two areas are about the same, but the mean for the Cleanup area is much larger, indicating a few or more "outlying" observations with large values. This may be indicative of residual contamination that was missed during the cleanup process.

Menu

To produce summary statistics for the original TcCB data using the ENVIRONMENTALSTATS pull-down menu, follow these steps.

1. Open the Object Explorer. If **epa.94b.tccb.df** is not visible in the Object Column, click on the **Find S-PLUS Objects** button (the binoculars icon) and in the Pattern box type **epa.94b.tccb.df**, then click **OK**.
2. Highlight the shortcut **epa.94b.tccb.df** in the Object column of the Object Explorer.
3. On the S-PLUS menu bar, make the following menu choices: **EnvironmentalStats>EDA>Summary Statistics**. This will bring up the Full Summary Statistics dialog box.
4. In the Data Set box, select or type **epa.94b.tccb.df** is selected. In the Variable(s) box, choose **TcCB**. In the Grouping Variables box, select **Area**.
5. Click **OK** or **Apply**.

Command

To produce summary statistics for the original TcCB data using the S-PLUS Command or Script Window, type this command:

```
> full.summary(split(TcCB, Area))
```

1.11.3 Looking at the TcCB Data with Histograms and Boxplots

Figure 1.1 shows the histograms for the TcCB concentrations in the Reference and Cleanup areas. Figure 1.2 shows side-by-side boxplots comparing the distribution of TcCB concentrations in the two areas. Both graphs display concentrations on the log scale. We see in these plots that most of the observations in the Cleanup area are comparable to (or even smaller than) the observations in the Reference area, but there are a few very large "outliers" in the Cleanup area.

Menu

To produce the histograms and boxplots for the log-transformed TcCB data using the S-PLUS pull-down menu, it will simplify things if we first make a new data frame called new.epa.94b.tccb.df to contain the original data and the log-transformed TcCB observations. To create the data frame new.epa.94b.tccb.df, follow these steps.

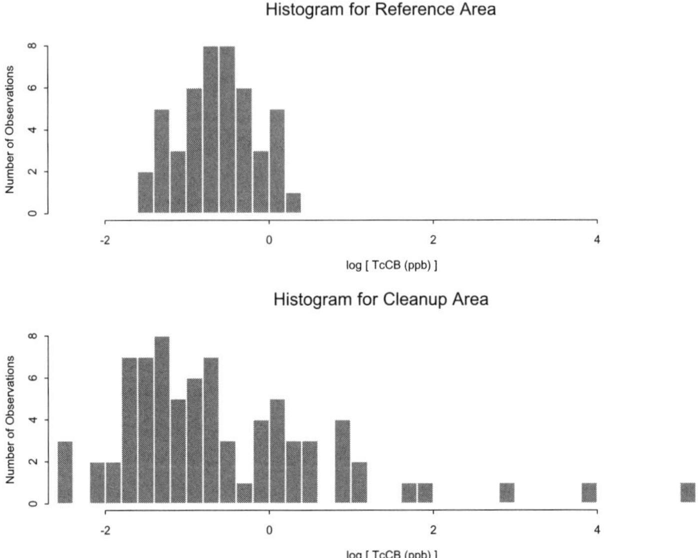

Figure 1.1. Histograms comparing TcCB concentrations at Reference and Cleanup areas.

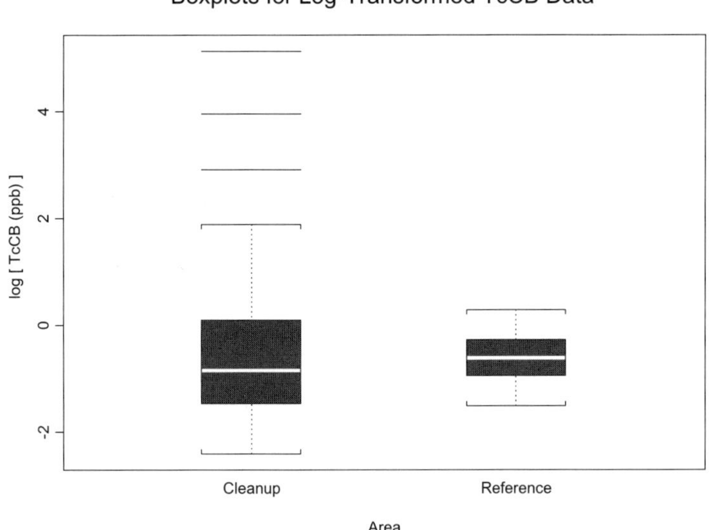

Figure 1.2. Boxplots comparing TcCB concentrations at Reference and Cleanup areas.

1. Highlight the shortcut **epa.94b.tccb.df** in the Object column of the Object Explorer.
2. On the S-PLUS menu bar, make the following menu choices: **Data>Transform**. This will bring up the Transform dialog box.
3. In the Target Column box, type **log.TcCB**. In the Variable box, choose **TcCB**. In the Function box choose **log**.
4. Click on the **Add** button, then click **OK**. At this point, you will get a warning message telling you that you have created a new copy of the data frame `epa.94b.tccb.df` that masks the original copy. Close the message window. Also, the modified data frame pops up in a data window. Close the data window.
5. In the left-hand column of the Object Explorer, click on the **Data** folder. In the right-hand column of the Object Explorer, right-click on **epa.94b.tccb.df** and choose **Properties**. In the Name box rename this data frame to **new.epa.94b.tccb.df** and click **OK**.

To produce the histograms shown in Figure 1.1 using the S-PLUS pull-down menus or toolbars, follow these steps.

1. In the Object Explorer, highlight the **Data** folder in the left-hand column. In the right-hand (Object) column, left-double-click the shortcut **new.epa.94b.tccb.df**. This will bring up a data window.
2. On the S-PLUS menu bar, make the following menu choices: **Graph>2D Plot**. This will bring up the Insert Graph dialog box. Under the Axes Type column, **Linear** should be highlighted. Under the Plot Type column, select **Histogram** and click **OK**. (Alternatively, left-click on the **2D** Plots button, then left-click on the **Histogram** button.)
3. The Histogram/Density dialog box should appear. Under the Data Columns group, in the Data Set box, select **new.epa.94b.tccb.df**. Under the Data Columns group, in the x Columns box, choose **log.TcCB**.
4. Click on the **Options** tab. Under Histogram Specs, in the Output Type box, select **Density**. For Lower Bound type **–2.5**. For Interval Width type **0.5**.
5. Click **OK**. A histogram of the data for both areas is displayed in a graphsheet.
6. **Click** on the data window to bring it forward. **Left click** on the top of the Area column to highlight that column, then **left-click** in a cell of the column, then **drag** the column to the top of the graphsheet and **drop** it.

To produce the boxplots shown in Figure 1.2 using the S-PLUS pull-down menus or toolbars, follow these steps.

1. On the S-PLUS menu bar, make the following menu choices: **Graph>2D Plot**. This will bring up the Insert Graph dialog box. Under the Axes Type column, **Linear** should be highlighted. Under the Plot Type column, select **Boxplot** and click **OK**. (Alternatively, left-click on the **2D** Plots button, then left-click on the **Box** button.)

   2.   The Boxplot dialog box should appear. The Data Set box should display **new.epa.94b.tccb.df**. In the x Columns box, choose **Area**. In the y Columns box, choose **log.TcCB**.

   3.   Click **OK**.

Command

To produce the histograms shown in Figure 1.1 using the S-PLUS Command or Script Window, type these commands.

```
> par(mfrow=c(2,1))
> hist(log(TcCB[Area=="Reference"]),
    xlim=range(log(TcCB)), xlab="log [ TcCB (ppb) ]",
    ylab="Number of Observations",
    main="Histogram for Reference Area")
> hist(log(TcCB[Area=="Cleanup"]), xlim=range(log(TcCB)),
    nclass=25, xlab="log [ TcCB (ppb) ]",
    ylab="Number of Observations",
    main="Histogram for Cleanup Area")
```

To produce the boxplots shown in Figure 1.2 using the S-PLUS Command or Script Window, type these commands:

```
> boxplot(split(log(TcCB), Area), xlab="Area",
    ylab="log [ TcCB (ppb) ]",
    main="Boxplots for Log-Transformed TcCB Data")
```

1.11.4 Quantile (Empirical CDF) Plots

Figure 1.3 shows the quantile plot, also called the empirical cumulative distribution function (cdf) plot, for the Reference area TcCB data. You can easily pick out the median as about 0.55 ppb and the quartiles as about 0.4 ppb and 0.75 ppb (compare these numbers to the ones listed on page 8). You can also see that the quantile plot quickly rises, then pretty much levels off after about 0.8 ppb, which indicates that the data are skewed to the right (see the histogram for the Reference area data in Figure 1.1).

Figure 1.4 shows the quantile plot with a fitted lognormal distribution. We see that the lognormal distribution appears to fit these data quite well. Figure 1.5 compares the empirical cdf for the Reference area with the empirical cdf for the Cleanup area for the log-transformed TcCB data. As we saw with the histograms and boxplots, the Cleanup area has quite a few extreme values compared to the Reference area.

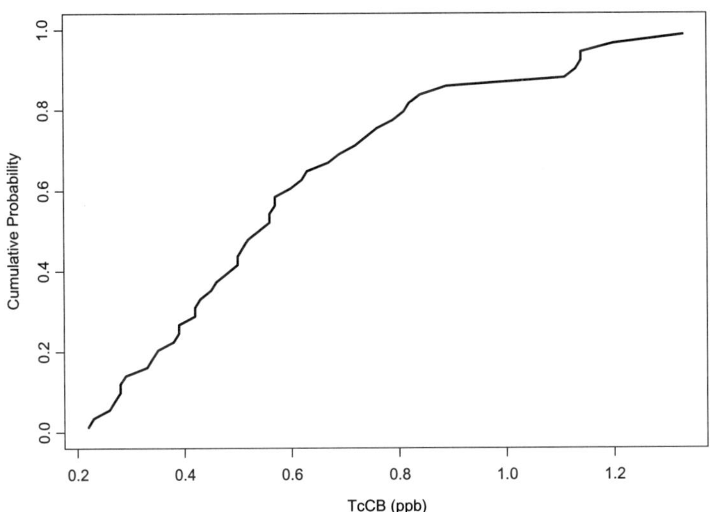

Figure 1.3. Quantile plot of Reference area TcCB data.

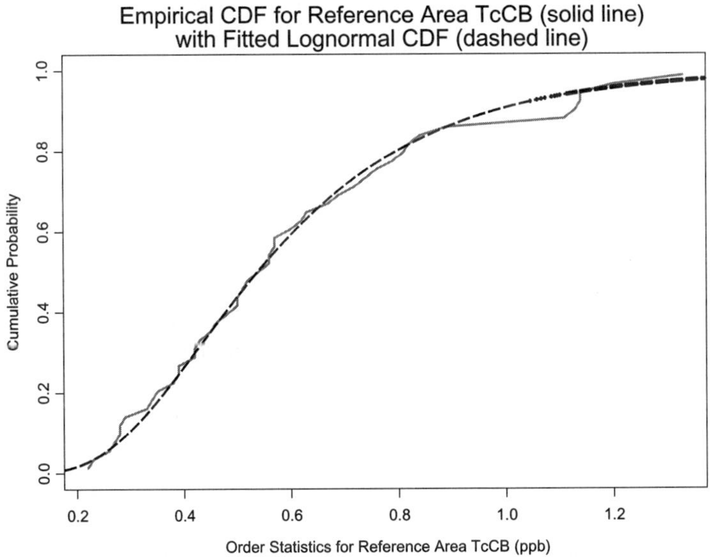

Figure 1.4. Empirical cdf of Reference area TcCB data compared to a lognormal cdf.

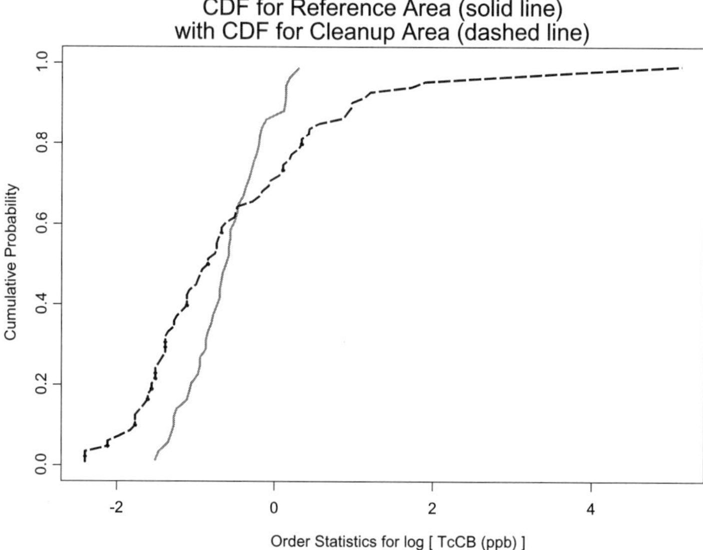

Figure 1.5. Quantile plots comparing log-transformed TcCB data at the Reference and Cleanup areas.

Menu

To produce the quantile plot of the Reference area TcCB data shown in Figure 1.3 using the ENVIRONMENTALSTATS for S-PLUS pull-down menu, follow these steps.

1. In the Object Explorer, highlight the **Data** folder in the left-hand column. In the right-hand (Object) column, highlight the shortcut **new.epa.94b.tccb.df** for the version of this data frame that you created earlier.

2. On the S-PLUS menu bar, make the following menu choices: **EnvironmentalStats>EDA>CDF Plot>Empirical CDF**. This will bring up the Plot Empirical CDF dialog box.

3. The Data Set box should display **new.epa.94b.tccb.df**. In the Variable(s) box, choose **TcCB**. In the Subset Rows with box, type **Area=="Reference"**.

4. Click **OK**.

To produce the quantile plot shown in Figure 1.4 comparing the Reference area TcCB data to a lognormal distribution, follow these steps.

1. On the S-PLUS menu bar, make the following menu choices: **EnvironmentalStats>EDA>CDF Plot>Compare Two CDFs**. This will bring up the Compare Two CDFs dialog box.

2. The Data Set box should display **new.epa.94b.tccb.df**. In the x Variable box, choose **TcCB**. In the Subset Rows with box, type **Area=="Reference"**. Under the Distribution Information Group, make sure that the **Estimate Parameters** box is checked. In the Distribution box, select **Lognormal**.
3. Click **OK** or **Apply**.

To produce the quantile plot shown in Figure 1.5 comparing the log-transformed Reference area TcCB data to the log-transformed Cleanup area TcCB data, follow these steps.

1. On the S-PLUS menu bar, make the following menu choices: **EnvironmentalStats>EDA>CDF Plot>Compare Two CDFs**. This will bring up the Compare Two CDFs dialog box.
2. For the Compare Data to choice, select **Other Data**. The Data Set box should display **new.epa.94b.tccb.df**. In the Variable 1 box, choose **log.TcCB**. In the Variable 2 box, choose **Area**. Click on the **Variable 2 is a Grouping Variable** box to check it.
3. Click **OK** or **Apply**.

Command

To produce the quantile plot of the Reference area TcCB data shown in Figure 1.3, type the following command.

```
> ecdfplot(TcCB[Area=="Reference"], xlab="TcCB (ppb)",
    main="Quantile Plot for Reference Area TcCB Data")
```

To produce the quantile plot shown in Figure 1.4 comparing the Reference area TcCB data to a lognormal distribution, type the following command.

```
> cdf.compare(TcCB[Area=="Reference"], dist="lnorm",
    xlab="Order Statistics for Reference Area TcCB
    (ppb)", main=paste("Empirical CDF for ",
      "Reference Area TcCB (solid line)\n",
      "with Fitted Lognormal CDF (dashed line)", sep=""))
```

To produce the quantile plot shown in Figure 1.5 comparing the log-transformed Reference area TcCB data to the log-transformed Cleanup area TcCB data, type the following command.

```
> cdf.compare(log(TcCB[Area=="Reference"]),
    log(TcCB[Area=="Cleanup"]),
    xlab="Order Statistics for log [ TcCB (ppb) ]",
    main=paste("CDF for Reference Area (solid line)",
      "with CDF for Cleanup Area (dashed line)",
      sep="\n"))
```

1.11.5 Assessing Goodness-of-Fit with Quantile-Quantile Plots

Figure 1.6 displays the normal Q-Q plot for the log-transformed Reference area TcCB data (i.e., we are assuming these data come from a lognormal distribution), along with a fitted least squares line. Figure 1.7 displays the corresponding Tukey mean-difference Q-Q plot. As we saw with the quantile plot, the lognormal model appears to be a fairly good fit to these data.

Menu

To create the normal Q-Q plot for the log-transformed Reference area data shown in Figure 1.6, follow these steps.

1. In the Object Explorer, highlight the **Data** folder in the left-hand column. In the right-hand (Object) column, highlight the shortcut **epa.94b.tccb.df**.
2. On the S-PLUS menu bar, make the following menu choices: **EnvironmentalStats>EDA>Q-Q Plot>Q-Q Plot**. This will bring up the Q-Q Plot dialog box.
3. The Data Set box should display **epa.94b.tccb.df**. In the x Variable box, choose **TcCB**. In the Subset Rows with box, type **Area=="Reference"**. In the Distribution box, make sure **Lognormal** is selected.
4. Click on the **Plotting** tab. Click on the **Add a Line** box to select this option.
5. Click **OK** or **Apply**.

To create the Tukey mean-difference Q-Q plot for the Reference area TcCB data fitted to a normal distribution shown in Figure 1.7, follow the same steps as above, but in Step 3 also check the **Estimate Parameters** box, and in Step 4 under Plot Type select **Tukey M-D**.

Command

To produce the Q-Q plot of the Reference area TcCB data fitted to a lognormal distribution shown in Figure 1.6, type the following command.

```
> qqplot(TcCB[Area=="Reference"], dist="lnorm",
     add.line=T, ylab="Quantiles of log [ TcCB (ppb) ]",
     main=paste("Normal Q-Q Plot for",
        "Log-Transformed Reference Area TcCB Data"))
```

To create the Tukey mean-difference Q-Q plot for the Reference area data fitted to a lognormal distribution shown in Figure 1.7, type the following command.

```
> qqplot(TcCB[Area=="Reference"], dist="lnorm",
     plot.type="Tukey", estimate.params=T, add.line=T,
     main=paste("Tukey Mean-Difference Q-Q Plot for",
```

```
            "\nReference Area TcCB Fitted to ",
            "Lognormal Distribution", sep=""))
```

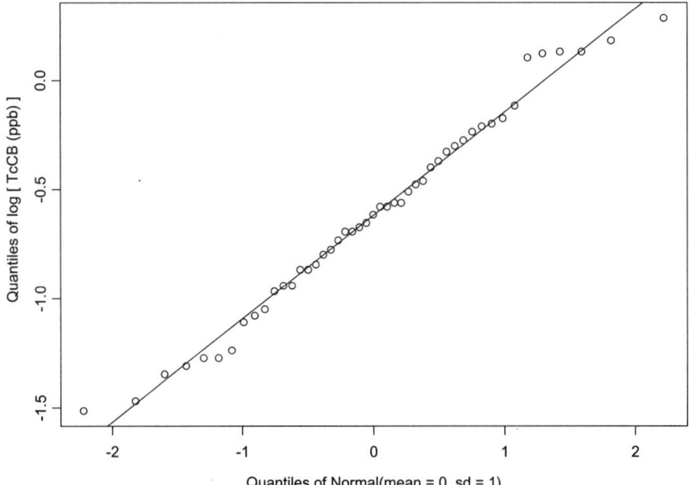

Figure 1.6. Normal Q-Q plot for the log-transformed Reference area TcCB data.

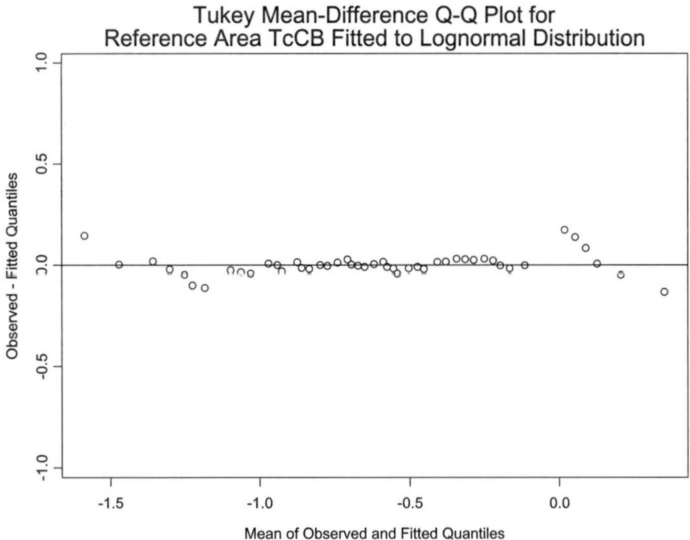

Figure 1.7. Tukey mean-difference Q-Q plot for the Reference area TcCB data fitted to a lognormal distribution.

1.11.6 Estimating Distribution Parameters

In ENVIRONMENTALSTATS for S-PLUS you can estimate parameters for several parametric distributions. For example, for the lognormal distribution, you can estimate the mean and standard deviation based on the log-transformed data, or you can estimate the mean and coefficient of variation based on the original data. For either parameterization, you can compute a confidence interval for the mean. Here are the results for the log-transformed Reference area TcCB data:

```
Results of Distribution Parameter Estimation
--------------------------------------------

Assumed Distribution:            Lognormal

Estimated Parameter(s):          meanlog = -0.6195712
                                 sdlog   =  0.467953

Estimation Method:               mvue

Data:                            TcCB[Area=="Reference"]

Sample Size:                     47

Confidence Interval for:         meanlog

Confidence Interval Method:      Exact

Confidence Interval Type:        two-sided

Confidence Level:                95%

Confidence Interval:             LCL = -0.7569673
                                 UCL = -0.4821751
```

and here are the results for the original Reference area TcCB data:

```
Results of Distribution Parameter Estimation
--------------------------------------------

Assumed Distribution:            Lognormal

Estimated Parameter(s):          mean = 0.5989072
                                 cv   = 0.4899539

Estimation Method:               mvue

Data:                            TcCB[Area=="Reference"]

Sample Size:                     47

Confidence Interval for:         mean
```

```
Confidence Interval Method:        Land

Confidence Interval Type:          two-sided

Confidence Level:                  95%

Confidence Interval:               LCL = 0.5243786
                                   UCL = 0.7016992
```

Menu

To compute the estimated mean and standard deviation for the log-transformed
Reference area TcCB data, follow these steps.

1. On the S-PLUS menu bar, make the following menu choices:
   EnvironmentalStats>Estimation>Parameters. This will bring up the
   Estimate Distribution Parameters dialog box.
2. For Data to Use, make sure the **Pre-Defined Data** button is selected.
   In the Data Set box, select or type **epa.94b.tccb.df**. In the Variable
   box, select **TcCB**. In the Subset Rows with box, type
   Area=="Reference".
3. In the Distribution/Estimation section, in the Distribution box select
   Lognormal, and for Estimation Method select **mvue**.
4. Under the Confidence Interval section, **check** the Confidence Interval
   box. In the CI Type box, select **two-sided**. In the CI Method box, select **exact**. In the Confidence Level (%) box, select **95**.
5. Click **OK** or **Apply**.

To compute the estimated mean and coefficient of variation for the original
Reference area TcCB data, follow the same steps above, but in Step 3 for Distribution select **Lognormal (Alternative)** and in Step 4 for CI Method select **land**.

Command

To compute the estimated mean and standard deviation for the log-transformed
Reference area TcCB data, type this command:

```
> elnorm(TcCB[Area=="Reference"], ci=T)
```

To compute the estimated mean and coefficient of variation for the original Reference area TcCB data, type this command:

```
> elnorm.alt(TcCB[Area=="Reference"], ci=T)
```

1.11.7 Testing for Goodness of Fit

ENVIRONMENTALSTATS for S-PLUS contains several new or modified S-PLUS
menu items and functions for testing goodness of fit. Here we will use the
Shapiro-Wilk test to test the adequacy of a lognormal model for the Reference
area TcCB data.

```
Results of Goodness-of-Fit Test
-------------------------------

Test Method:                     Shapiro-Wilk GOF

Hypothesized Distribution:       Lognormal

Estimated Parameter(s):          meanlog = -0.6195712
                                 sdlog   =  0.467953

Estimation Method:               mvue

Data:                            TcCB.ref

Sample Size:                     47

Test Statistic:                  W = 0.9789915

Test Statistic Parameter:        n = 47

P-value:                         0.5512532

Alternative Hypothesis:          True cdf does not equal
                                 Lognormal Distribution.
```

ENVIRONMENTALSTATS for S-PLUS also contains a plotting method for the results of goodness-of-fit tests, as well as a function or menu choice to produce four summary plots on one page, as shown in Figure 1.8.

Menu

To perform the Shapiro-Wilk goodness-of-fit test for lognormality on the Reference area TcCB data and produce a summary plot of the results, follow these steps.

1. On the S-PLUS menu bar, make the following menu choices: **EnvironmentalStats>Hypothesis Tests>GOF Tests>One Sample>Shapiro-Wilk**. This will bring up the Shapiro-Wilk GOF Test dialog box.
2. For Data to Use, select **Pre-Defined Data**. For Data Set make sure **epa.94b.tccb.df** is selected, for Variable select **TcCB**, and in the Subset Rows with box type **Area=="Reference"**. In the Distribution box select **Lognormal**.
3. Click on the **Plotting** tab. Under Plotting Information, in the Significant Digits box select **3**, then click **OK** or **Apply**.

Command

To perform the Shapiro-Wilk goodness-of-fit test for lognormality on the Reference area TcCB data, type these commands.

```
> TcCB.ref <- TcCB[Area=="Reference"]
> sw.list <- sw.gof(TcCB.ref, dist="lnorm")
> sw.list
```

To produce a summary plot of the results, type this command.

```
> plot.gof.summary(sw.list, digits=3)
```

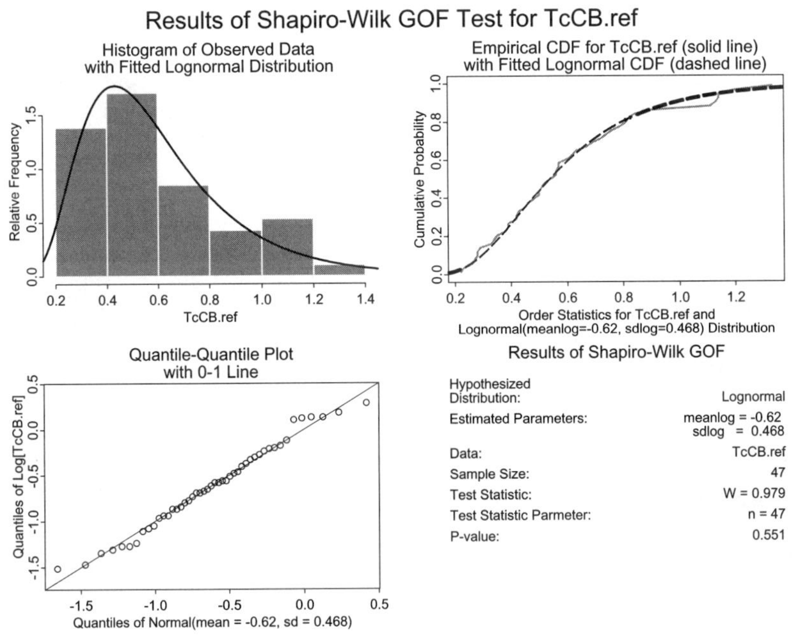

Figure 1.8. Summary plots of Shapiro-Wilk goodness-of-fit test for Reference area TcCB data.

1.11.8 Estimating Quantiles and Computing Confidence Limits

ENVIRONMENTALSTATS for S-PLUS contains a menu item and functions for estimating quantiles and optionally constructing confidence limits for the quantiles. Here we will estimate the 90th percentile of the distribution of the Reference area TcCB, assuming the true distribution is a lognormal distribution, and compute a 95% confidence interval for this 90th percentile.

Menu

To estimate the 90th percentile and compute a two-sided 95% confidence interval for this percentile, follow these steps.

1. On the S-PLUS menu bar, make the following menu choices: **EnvironmentalStats>Estimation>Quantiles**. This will bring up the Estimate Distribution Quantiles dialog box.
2. For Data to Use, make sure the **Pre-Defined Data** button is selected. In the Data Set box, select **epa.94b.tccb.df**. In the Variable box, select **TcCB**. In the Subset Rows with box, type **Area=="Reference"**.
3. In the Quantile(s) box, select **0.90**.
4. Under the Distribution/Estimation group, for Type, select the **Parametric** button. In the Distribution box, select **Lognormal**. For Estimation Method, select **qmle**.
5. Under the Confidence Interval group, make sure the Confidence Interval box is **checked**. For CI Type, select **two-sided**. For CI Method, select **exact**. For Confidence Level (%), select **95**.
6. Click **OK** or **Apply**.

Command

To estimate the 90th percentile and compute a two-sided 95% confidence interval for this percentile, type this command.

```
> eqlnorm(TcCB[Area=="Reference"], p=0.9, ci=T)

Results of Distribution Parameter Estimation
--------------------------------------------------

Assumed Distribution:            Lognormal

Estimated Parameter(s):          meanlog = -0.6195712
                                 sdlog   =  0.467953

Estimation Method:               mvue

Estimated Quantile(s):           90'th %ile = 0.9803307

Quantile Estimation Method:      qmle

Data:                            TcCB[Area=="Reference"]

Sample Size:                     47

Confidence Interval for:         90'th %ile

Confidence Interval Method:      Exact

Confidence Interval Type:        two-sided

Confidence Level:                95%

Confidence Interval:             LCL = 0.8358791
                                 UCL = 1.215498
```

1.11.9 Comparing Two Distributions Using Nonparametric Tests

ENVIRONMENTALSTATS for S-PLUS contains menu items and functions for performing general two-sample linear rank tests (to test for a shift in location) and a special quantile test that tests for a shift in the tail of one of the distributions. In this example we will compare the Reference and Cleanup area TcCB data. Here are the results for the Wilcoxon rank sum test.

```
Results of Hypothesis Test
--------------------------


Null Hypothesis:          Fy(t) = Fx(t)

Alternative Hypothesis:   Fy(t) > Fx(t) for at least one t

Test Name:                Two-Sample Linear Rank Test:
                          Wilcoxon Rank Sum Test
                          Based on Normal Approximation

Data:                     x = TcCB[Area == "Cleanup"]
                          y = TcCB[Area == "Reference"]

Sample Sizes:             nx = 77
                          ny = 47

Test Statistic:           z = -1.171872

P-value:                  0.8793758
```

and here are the results for the quantile test

```
Results of Hypothesis Test
--------------------------


Null Hypothesis:          e = 0

Alternative Hypothesis:   Tail of Fx Shifted to Right of
                          Tail of Fy.
                          0 < e <= 1, where
                          Fx(t) = (1-e)*Fy(t) + e*Fz(t),
                          Fz(t) <= Fy(t) for all t,
                          and Fy != Fz

Test Name:                Quantile Test

Data:                     x = TcCB[Area=="Cleanup"]
                          y = TcCB[Area=="Reference"]

Sample Sizes:             nx = 77
                          ny = 47
```

```
Test Statistics:                    k (# x obs of r largest) = 9
                                    r                         = 9

Test Statistic Parameters:          m              = 77
                                    n              = 47
                                    quantile.ub =   0.928

P-value:                            0.01136926
```

Note that the Wilcoxon rank sum test is not significant at the 0.10 level (p=0.88), while the quantile test is significant at the 0.011 level. The quantile test picked up the portion of large outlying values in the Cleanup area data.

Menu

To perform the Wilcoxon rank sum test, follow these steps.

1. On the S-PLUS menu bar, make the following menu choices: **EnvironmentalStats>Hypothesis Tests>Compare Samples>Two Samples>Linear Rank Test**. This will bring up the Two-Sample Linear Rank Test dialog box.
2. For Data Set, select **epa.94b.tccb.df**. For Variable 1 select **TcCB**, for Variable 2 select **Area**, and **check** the box that says Variable 2 is a Grouping Variable.
3. For Test select **Wilcoxon**, and for Shift Type select **Location**. Under Alternative Hypothesis select **greater**, then click **OK** or **Apply**.

To perform the quantile test, follow these steps.

1. On the S-PLUS menu bar, make the following menu choices: **EnvironmentalStats>Hypothesis Tests>Compare Samples>Two Samples>Quantile Test**. This will bring up the Two-Sample Quantile Test dialog box.
2. For Data Set, select **epa.94b.tccb.df**. For Variable 1 select **TcCB**, for Variable 2 select **Area**, and **check** the box that says Variable 2 is a Grouping Variable.
3. Under Alternative Hypothesis select **greater**. For Specify Target select **Rank**, and for Target Rank type **9**, then click **OK** or **Apply**.

Command

To perform the Wilcoxon rank sum test, type this command

```
> wilcox.test(TcCB[Area=="Cleanup"],
    TcCB[Area=="Reference"], alternative="greater")
```

or this command

```
> two.sample.linear.rank.test(TcCB[Area=="Cleanup"],
    TcCB[Area=="Reference"], alternative="greater")
```

Note that by default the S-PLUS function `wilcox.test` computes the "corrected" z-statistic while the ENVIRONMENTALSTATS for S-PLUS function `two.sample.linear.rank.test` computes the "uncorrected" z-statistic.

To perform the quantile test, type this command.

```
> quantile.test(TcCB[Area=="Cleanup"],
    TcCB[Area=="Reference"], alternative="greater",
    target.r=9)
```

1.12 Summary

- Environmental statistics is the application of statistics to environmental problems.
- ENVIRONMENTALSTATS for S-PLUS is an S-PLUS module for environmental statistics. It includes several menu items and functions for creating graphs and performing statistical analyses that are commonly used in environmental statistics.
- To use ENVIRONMENTALSTATS for S-PLUS you should be familiar with the basic operation of S-PLUS and have an elementary knowledge of probability and statistics.
- ENVIRONMENTALSTATS for S-PLUS has an extensive help system that includes basic explanations in English, as well as equations and references.

2

Designing a Sampling Program

2.1 Introduction

The first and most important step of any environmental study is to design the sampling program. This chapter discusses the basics of designing a sampling program, and shows you how to use ENVIRONMENTALSTATS for S-PLUS to help you determine required sample sizes. For a more in-depth discussion of sampling design and sample size calculation, see chapters 2 and 8 of Millard and Neerchal (2001).

2.2 The Necessity of a Good Sampling Design

A study is only as good as the data upon which it is based. No amount of advanced, cutting-edge statistical theory and techniques can rescue a study that has produced poor quality data, not enough data, or data irrelevant to the question it was meant to answer. From the very start of an environmental study, there must be a constant dialog between the data producers (field and lab personnel, data coders, etc.), the data users (scientists and statisticians), and the ultimate decision maker (the person for whom the study was instigated in the first place). All persons involved in the study must have a clear understanding of the study objectives and the limitations associated with the chosen physical sampling and analytical (measurement) techniques before anyone can make any sense of the resulting data.

2.3 What is a Population and What is a Sample?

In everyday language, the word "population" refers to all the people or organisms contained within a specific country, area, region, etc. When we talk about the population of the United States, we usually mean something like "the total number of people who currently reside in the U.S."

In the field of statistics, however, the term *population* is defined operationally by the question we ask: it is the entire collection of measurements about which we want to make a statement (Zar, 1999, p.16; Berthoux and Brown, 1994, p.7; Gilbert, 1987, Chapter 2).

For example, if the question is "What is concentration of dissolved oxygen in this stream?", the question must be further refined until a suitable population can be defined: "What is the average concentration of dissolved oxygen in a particular section of a stream at a depth of 0.5 m over a particular 3-day period?" In this case, the population is the set of all possible measurements of dissolved oxygen in that section of the stream at 0.5 m within that time period. The section of the stream, the time period, the method of taking water samples, and the method of measuring dissolved oxygen all define the population.

A *sample* is defined as some subset of a population (Zar, 1999, p.17; Berthoux and Brown, 1994, p.7; Gilbert, 1987, Chapter 2). If the sample contains all the elements of the population, it is called a *census*. Usually, a population is too large to take a census, so a portion of the population is sampled. The statistical definition of the word sample (a selection of individual population members) should not be confused with the more common meaning of a *physical sample* of soil (e.g., 10g of soil), water (e.g., 5ml of water), air (e.g., 20 cc of air), etc.

2.4 Random versus Judgment Sampling

Judgment sampling involves subjective selection of the population units by an individual or group of individuals (Gilbert, 1987, Chapter 3). For example, the number of samples and sampling locations might be determined based on expert opinion or historical information. Sometimes, public opinion might play a role and samples need to be collected from areas known to be highly polluted. The uncertainty inherent in the results of a judgment sample cannot be quantified and statistical methods cannot be applied to judgment samples. Judgment sampling does *not* refer to using prior information and the knowledge of experts to define the area of concern, define the population, or plan the study. Gilbert (1987, p. 19) also describes "haphazard" sampling, which is a kind of judgment sampling with the attitude that "any sample will do" and can lead to "convenience" sampling, in which samples are taken in convenient places at convenient times.

Probability sampling or *random sampling* involves using a random mechanism to select samples from the population (Gilbert, 1987, Chapter 3). All statistical methods used to quantify uncertainty assume some form of random sampling has been used to obtain a sample. At the simplest level, a simple random sample is used in which each member of the population has an equal chance of being chosen, and the selection of any member of the population does not influence the selection of any other member. Other probability sampling methods include stratified random sampling, composite sampling, and ranked set sampling (see the help file *Glossary: Sample and Sampling Methods* for more information).

2.5 Common Mistakes in Environmental Studies

The most common mistakes that occur in environmental studies include the following:

- **Lack of Samples from Proper Control Populations**. If one of the objectives of an environmental study is to determine the effects of a pollutant on some specified population, then the sampling design must include samples from a proper control population. This is a basic tenet of the scientific method. If control populations were not sampled, there is no way to know whether the observed effect was really due to the hypothesized cause, or whether it would have occurred anyway.

- **Using Judgment Sampling to Obtain Samples.** When judgment sampling is used to obtain samples, there is no way to quantify the precision and bias of any type of estimate computed from these samples.

- **Failing to Randomize over Potentially Influential Factors.** An enormous number of factors can influence the final measure associated with a single sampling unit, including the person doing the sampling, the device used to collect the sample, the weather and field conditions when the sample was collected, the method used to analyze the sample, the laboratory to which the sample was sent, etc. A good sampling design controls for as many potentially influencing factors as possible, and randomizes over the factors that cannot be controlled. For example, if there are four persons who collect data in the field, and two laboratories are used to analyze the results, you should not send all the samples collected by persons 1 and 2 to laboratory A and all the samples collected by persons 3 and 4 to laboratory B, but rather send samples collected by each person to each of the laboratories.

- **Collecting Too Few Samples to Have a High Degree of Confidence in the Results.** The ultimate goal of an environmental study is to answer one or more basic questions. These questions should be stated in terms of hypotheses that can be tested using statistical procedures. In this case, you can determine the probability of rejecting the null hypothesis when in fact it is true (a Type I error), and the probability of not rejecting the null hypothesis when if fact it is false (a Type II error). Usually, the Type I error is set in advance, and the probability of correctly rejecting the null hypothesis when in fact it is false (the power) is calculated for various sample sizes. Too often, this step of determining power and sample size is neglected, resulting in a study from which no conclusions can be drawn with any great degree of confidence.

2.6 The Data Quality Objectives Process

The Data Quality Objectives (DQO) process is a systematic planning tool based on the scientific method that has been developed by the U.S. Environmental Protection Agency (USEPA, 1994a). The DQO process provides an easy-to-follow, step-by-step approach to decision-making in the face of uncertainty. Each step focuses on a specific aspect of the decision-making process. Data Quality Objectives are the qualitative and quantitative statements that:

- Clarify the study objective.
- Define the most appropriate type of data to collect.
- Determine the most appropriate conditions under which to collect the data.
- Specify acceptable levels of decision errors that will be used as the basis for establishing the quantity and quality of data needed to support the decision.

The seven steps in the DQO process are: state the problem, identify the decision, identify inputs to the decision, define the study boundaries, develop a decision rule, specify tolerable limits on decision errors, and optimize the design (see Chapter 2 of Millard and Neerchal, 2001, for more details). These last two steps involve trading off limits on Type I and Type II errors and sample size.

2.7 Power and Sample Size Calculations

ENVIRONMENTALSTATS for S-PLUS contains several menu items and functions to assist you in determining how many samples you need for a given degree of confidence in the results of a sampling program (see the help file *Power and Sample Size Calculations*). These menu items and functions are based on the ideas of confidence intervals, prediction intervals, tolerance intervals, and hypothesis tests. If you are unfamiliar with these concepts, please see chapters 5–7 of Millard and Neerchal (2001).

A very important point to remember is that no matter what you come up with for estimates of required sample sizes, it is always a good idea to assume you will lose some percentage of your observations (say 10%) due to sample loss, sample contamination, data coding errors, misplaced forms, etc.

2.8 Sample Size Calculations for Confidence Intervals

Table 2.1 lists the menu items and functions available in ENVIRONMENTALSTATS for S-PLUS for computing required sample sizes, half-widths, and confidence levels associated with a confidence interval. All menu items fall under the menu choice **EnvironmentalStats>Sample Size and Power>Confidence Intervals**. For example, the first entry in the table describes what you can do if you make the selection **Environmental-Stats>Sample Size and Power>Confidence Intervals>Normal Mean>Compute**.

Menu Item	Functions	Description
Normal Mean> Compute	`ci.norm.half.width`	Compute half-width of a confidence interval for mean of a normal distribution or difference between two means.
	`ci.norm.n`	Compute required sample size for specified half-width of a confidence interval for mean of a normal distribution or difference between two means.
Normal Mean> Plot	`plot.ci.norm.design`	Create plots for sampling design based on a confidence interval for mean of normal distribution or difference between two means.
Binomial Proportion> Compute	`ci.binom.half.width`	Compute half-width of a confidence interval for binomial proportion or difference between two proportions.
	`ci.binom.n`	Compute required sample size for specified half-width of a confidence interval for binomial proportion or difference between two proportions.
Binomial Proportion> Plot	`plot.ci.binom.design`	Create plots for sampling design based on a confidence interval for binomial proportion or difference between two proportions.
Nonparametric> Compute	`ci.npar.conf.level`	Compute confidence level of a confidence interval for a percentile, given the sample size.
	`ci.npar.n`	Compute required sample size for specified confidence level of a confidence interval for a percentile.
Nonparametric> Plot	`plot.ci.npar.design`	Create plots for sampling design based on a confidence interval for a percentile.

Table 2.1. Sample size menu items and functions for confidence intervals.

For the normal and binomial distributions, you can compute the half-width of the confidence interval given the user-specified sample size, compute the required sample size given the user-specified half-width, and plot the relationship between sample size and half-width. For a nonparametric confidence interval for a percentile, you can compute the required sample size for a specified confidence level, compute the confidence level associated with a given sample size, and plot the relationship between sample size and confidence level.

2.8.1   Confidence Interval for the Mean of a Normal Distribution

The help files for `ci.norm.half.width` and `ci.norm.n` explain how to compute a confidence interval for a mean of a normal distribution or the difference between two means, as well as the relationship between the sample size, estimated standard deviation, confidence level, and half-width. The function `ci.norm.half.width` computes the half-width associated with the confidence interval, given the sample size, estimated standard deviation, and confidence level. The function `ci.norm.n` computes the sample size required to achieve a specified half-width, given the estimated standard deviation and confidence level. The function `plot.ci.norm.design` plots the relationships between sample size, half-width, estimated standard deviation, and confidence level.

The data frame `sulfate.df` contains sulfate concentrations (ppm) at one background and one downgradient well. The estimated mean and standard deviation for the background well are 536 and 27 ppm, respectively, based on a sample size of $n = 8$ quarterly samples. A two-sided 95% confidence interval for this mean is [514, 559], which has a half-width of 22.5 ppm. Assuming a standard deviation of about $s = 30$, if we had taken only four observations, the half-width of the confidence interval would have been about 48 ppm. Also, if we want the confidence interval to have a half-width of 10 ppm, we would need to take $n = 38$ observations. Figure 2.1 displays the half-width of the confidence interval as a function of the sample size for various confidence levels.

Menu

To compute the half-width of the confidence interval assuming a sample size of $n = 4$ using the ENVIRONMENTALSTATS for S-PLUS pull-down menu, follow these steps.

1.   On the S-PLUS menu bar, make the following menu choices: **EnvironmentalStats>Sample Size and Power>Confidence Intervals>Normal Mean>Compute**. This will bring up the Normal CI Half-Width and Sample Size dialog box.
2.   For Compute, select **CI Half-Width**, for Sample Type select **One-Sample**, for Confidence Level(s) (%) select **95**, for n type **4**, and for Standard Deviation(s) type **30**. Click **OK** or **Apply**.

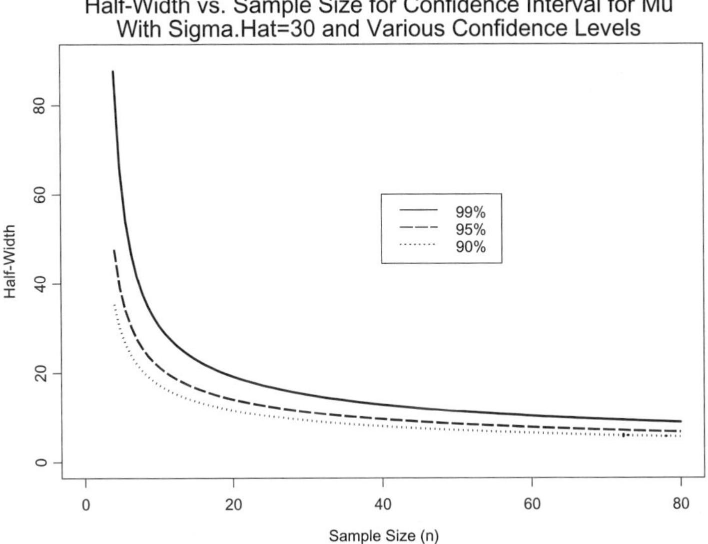

Figure 2.1. The half-width of the confidence interval for the mean of background sulfate concentration (ppm) as a function of sample size and confidence level.

To compute the required sample size for a specified half-width of 10, follow the same steps as above, except for Compute select **Sample Size**, and for Half-Width type **10**.

To create the plot shown in Figure 2.1, follow these steps.

1. On the S-PLUS menu bar, make the following menu choices: **EnvironmentalStats>Sample Size and Power>Confidence Intervals>Normal Mean>Plot**. This will bring up the Plot Normal CI Design dialog box.

2. For X Variable select **Sample Size**, for Y Variable select **Half-Width**, for Minimum X type **4**, for Maximum X type **80**, for Sample Type select **One-Sample**, for Confidence Level (%) select **99**, and for Standard Deviation type **30**.

3. Click on the **Plotting** tab. **Uncheck** the Add Title box. In the X-Axis Limits box type **0, 80**. In the Y-Axis Limits box type **0, 90**. Click **Apply**.

4. Click on the **Model** tab. For Confidence Level (%) select **95**. Click on the **Plotting** tab, **check** the Add to Current Plot box, and for Line Type select the fourth pattern. Click **Apply**.

5. Click on the **Model** tab. For Confidence Level (%) select **90**. Click on the **Plotting** tab, and for Line Type select the eighth pattern. Click **OK**.

Command

To compute the half-width of the confidence interval assuming a sample size of $n = 4$ using the ENVIRONMENTALSTATS for S-PLUS Command or Script Window, type this command.

```
> ci.norm.half.width(n.or.n1=4, sigma.hat=30)
```

To compute the required sample size for a specified half-width of 10, type this command.

```
> ci.norm.n(half.width=10, sigma.hat=30)
```

To create the plot shown in Figure 2.1, type these commands.

```
> plot.ci.norm.design(sigma.hat=30,
    range.x.var=c(4, 80), conf=0.99, xlim=c(0, 80),
    ylim=c(0, 90), main="Half-Width vs. Sample Size for
    Confidence Interval for Mu\nWith Sigma.Hat=30 and
    Various Confidence Levels")
> plot.ci.norm.design(sigma.hat=30, range.x.var=c(4, 80),
    conf=0.95, plot.lty=4, add=T)
> plot.ci.norm.design(sigma.hat=30, range.x.var=c(4, 80),
    conf=0.90, plot.lty=2, add=T)
> legend(40, 60, c("99%", "95%", "90%"),
    lty=c(1, 4, 2), lwd=1.5)
```

2.8.2   Confidence Interval for a Binomial Proportion

The help files for `ci.binom.half.width` and `ci.binom.n` explain how to compute a confidence interval for a binomial proportion or the difference between two proportions, as well as the relationship between the sample size, estimated proportions, confidence level, and half-width.   The function `ci.binom.half.width` computes the half-width associated with the confidence interval, given the sample size, estimated proportion(s), and confidence level.   The function `ci.binom.n` computes the sample size required to achieve a specified half-width, given the estimated proportion(s) and confidence level.   The function `plot.ci.binom.design` plots the relationships between sample size, half-width, estimated proportion(s), and confidence level.

The data frame `epa.92c.benzene1.df` contains observations on benzene concentrations (ppb) in groundwater from six background wells sampled monthly for 6 months. Nondetect values are reported as "<2." Of the 36 values, 33 are nondetects. Based on these data, the estimated probability of observing a nondetect is 92%, and the two-sided 95% confidence interval for the binomial proportion is [0.78, 0.98]. The half-width of this interval is 0.1, or 10 percentage points. Assuming an estimated proportion of 90%, if we had taken only $n = 10$ observations, the half-width of the confidence interval would have been about 0.22 (22 percentage points). Also, if we want the confidence interval to have a half-width of 0.02 (two percentage points), we would need to take $n =$ 914 observations. Figure 2.2 displays the half-width of the confidence interval as a function of the sample size for various confidence levels.

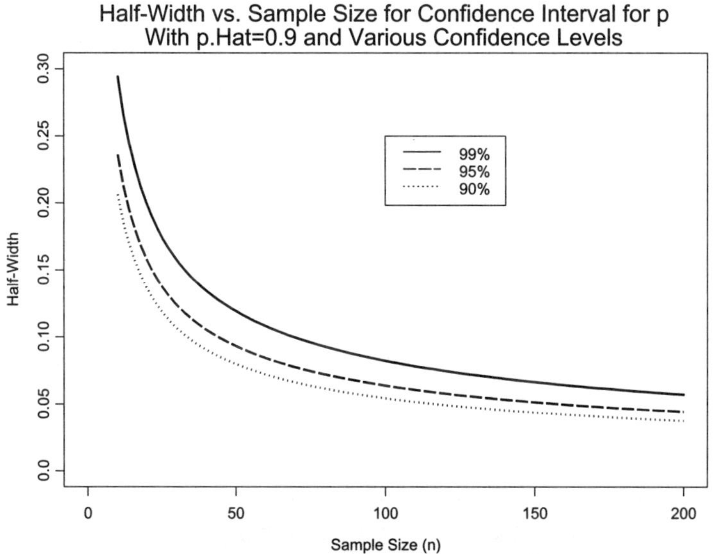

Figure 2.2. The half-width of the confidence interval for the probability of a nondetect as a function of sample size and confidence level, assuming an estimated nondetect proportion of about 90%.

Menu

To compute the half-width of the confidence interval assuming a sample size of $n = 10$ using the ENVIRONMENTALSTATS for S-PLUS pull-down menu, follow these steps.

1. On the S-PLUS menu bar, make the following menu choices: **EnvironmentalStats>Sample Size and Power>Confidence Intervals>Binomial Proportion>Compute**. This will bring up the Binomial CI Half-Width and Sample Size dialog box.
2. For Compute, select **CI Half-Width**, for Sample Type select **One-Sample**, for Confidence Level(s) (%) select **95**, for n type **10**, and for Proportion type **0.9**. Click **OK** or **Apply**.

To compute the required sample size for a specified half-width of 0.02, follow the same steps as above, except for Compute select **Sample Size**, and for Half-Width type **0.02**.

To create the plot shown in Figure 2.2, follow these steps.

1. On the S-PLUS menu bar, make the following menu choices: **EnvironmentalStats>Sample Size and Power>Confidence Intervals>Binomial Proportion>Plot**. This will bring up the Plot Binomial CI Design dialog box.
2. For X Variable select **Sample Size**, for Y Variable select **Half-Width**, for Minimum X type **10**, for Maximum X type **200**, for Sample Type select **One-Sample**, for Confidence Level (%) select **99**, and for Proportion type **0.9**.
3. Click on the **Plotting** tab. **Uncheck** the Add Title box. In the X-Axis Limits box type **0, 200**. In the Y-Axis Limits box type **0, 0.3**. Click **Apply**.
4. Click on the **Model** tab. For Confidence Level (%) select **95**. Click on the **Plotting** tab, **check** the Add to Current Plot box, and for Line Type select the fourth pattern. Click **Apply**.
5. Click on the **Model** tab. For Confidence Level (%) select **90**. Click on the **Plotting** tab, and for Line Type select the eighth pattern. Click **OK**.

Command

To compute the half-width of the confidence interval assuming a sample size of $n = 10$ using the ENVIRONMENTALSTATS for S-PLUS Command or Script Window, type this command.

```
> ci.binom.half.width(n.or.n1=10, p.hat=0.9, approx=F)
```

To compute the required sample size for a specified half-width of 0.02, type this command.

```
> ci.binom.n(half.width=0.2, p.hat=0.9)
```

To create the plot shown in Figure 2.2, type these commands.

```
> plot.ci.binom.design(p.hat=0.9, range.x.var=c(10, 200),
    conf=0.99, xlim=c(0, 200), ylim=c(0, 0.3),
    main="Half-Width vs. Sample Size for Confidence
    Interval for p\nWith p.Hat=0.9 and Various Confidence
    Levels")
> plot.ci.binom.design(p.hat=0.9, range.x.var=c(10, 200),
    conf=0.95, plot.lty=4, add=T)
> plot.ci.binom.design(p.hat=0.9, range.x.var=c(10, 200),
    conf=0.90, plot.lty=2, add=T)
> legend(100, 0.25, c("99%", "95%", "90%"),
    lty=c(1, 4, 2), lwd=1.5)
```

2.8.3   Nonparametric Confidence Interval for a Percentile

The help files for `ci.npar.conf.level` and `ci.npar.n` explain how to compute a nonparametric confidence interval for a quantile (the pth quantile is the same as the $100p^{th}$ percentile, where $0 \le p \le 1$), as well as the relationship between the quantile, sample size, and confidence level.   The function `ci.npar.conf.level` computes the confidence level associated with the confidence interval, given the sample size and value of p.   The function `ci.npar.n` computes the sample size required to achieve a specified confidence level, given the value of p.   The function `plot.ci.npar.design` plots the relationships between sample size, confidence level, and p.

The data frame `epa.92c.copper2.df` contains copper concentrations (ppb) at three background wells and two compliance wells. There are 8 observations associated with each of the three background wells. Of the 24 observations at the three background wells, 15 are nondetects recorded as "<5.". The other 9 observations at the background wells are: 5.4, 5.9, 6.0, 6.1, 6.4, 6.7, 7.5, 8.0, and 9.2. The estimated 95th percentile of copper concentration at the background wells is 7.925. If we use the largest observed value of 9.2 as the upper confidence limit of the 95th percentile of the copper concentration, the associated confidence level is 71%. If only 4 observations had been taken at each well for a total sample size of $n = 12$, the associated confidence level would have been 46%. If we want to construct a nonparametric confidence interval for the 95th percentile of copper concentration with an associated confidence level of at least 95%, we would need $n = 59$ observations (about 20 observations at each background well). Figure 2.3 displays the confidence level of the one-sided upper confidence interval for the 95th percentile as a function of the sample size.

Menu

To compute the confidence level of the one-sided upper confidence interval for the 95th percentile assuming a sample size of $n = 12$ using the ENVIRONMENTALSTATS for S-PLUS pull-down menu, follow these steps.

1.   On the S-PLUS menu bar, make the following menu choices: **EnvironmentalStats>Sample Size and Power>Confidence Intervals>Nonparametric>Compute**. This will bring up the Nonparametric CI Confidence Level and Sample Size dialog box.
2.   For Compute select **Confidence Level**, for n type **12**, for Percentile(s) select **95**, and for CI Type select **upper**. Click **OK** or **Apply**.

To compute the required sample sizes for a specified confidence level of 95%, follow the same steps as above, except for Compute select **Sample Size**, and for Confidence Level (%) select **95** (note that the box for n will be grayed out).

To create the plot shown in Figure 2.3, follow these steps.

1. On the S-PLUS menu bar, make the following menu choices: **EnvironmentalStats>Sample Size and Power>Confidence Intervals>Nonparametric>Plot**. This will bring up the Plot Nonparametric CI Design dialog box.

2. For X Variable select **Sample Size**, for Y Variable select **Confidence**, for Minimum X type **2**, for Maximum X type **100**, for Percentile select **95**, and for CI Type select **upper**. Click on the **Plotting** tab. In the Y-Axis Limits box type **0, 1**. Click **OK**.

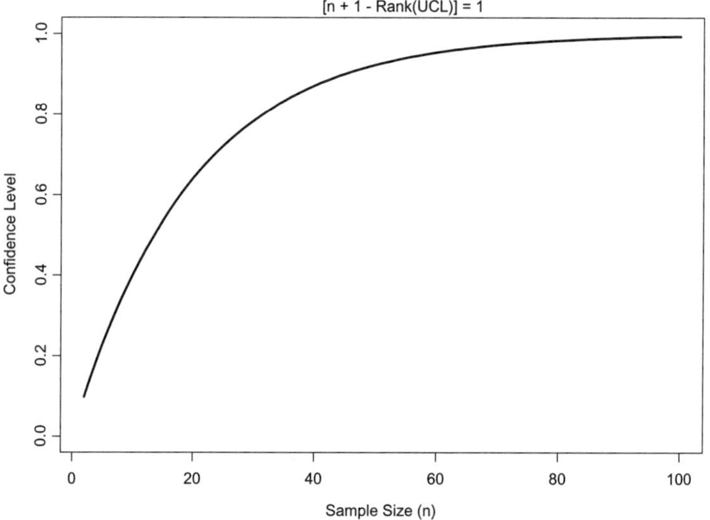

Figure 2.3. The confidence level of the one-sided upper nonparametric confidence interval for the 95th percentile.

Command

To compute the confidence level of the one-sided upper confidence interval for the 95th percentile assuming a sample size of $n = 12$ using the ENVIRONMENTALSTATS for S-PLUS Command or Script Window, type this command.

```
> ci.npar.conf.level(n=12, p=0.95, ci.type="upper")
```

To compute the required sample size for a specified confidence level of 95%, type this command.

```
> ci.npar.n(p=0.95, ci.type="upper")
```

To create the plot shown in Figure 2.3, type this command.

```
> plot.ci.npar.design(p=0.95, ci.type="upper",
    range.x.var=c(2, 100), ylim=c(0, 1))
```

2.9   Sample Size Calculations for Prediction Intervals

Table 2.2 lists the menu items and functions available in ENVIRONMENTALSTATS for S-PLUS for computing required sample sizes, half-widths, and confidence levels associated with a prediction interval. All menu items fall under the menu choice **EnvironmentalStats>Sample Size and Power>Prediction Intervals**. For example, the first entry in the table describes what you can do if you make the selection **EnvironmentalStats>Sample Size and Power>Prediction Intervals>Normal>Compute**.

Menu Item	**Functions**	**Description**
Normal> Compute	`pred.int.norm.half.width`	Compute half-width of a prediction interval for a normal distribution.
	`pred.int.norm.n`	Compute required sample size for specified half-width of a prediction interval for a normal distribution.
Normal> Plot	`plot.pred.int.norm.design`	Create plots for sampling design based on prediction interval for a normal distribution.
Nonparametric> Compute	`pred.int.npar.conf.level` `pred.int.npar.simultaneous.conf.level`	Compute confidence level of a prediction interval, given the sample size.
	`pred.int.npar.n` `pred.int.npar.simultaneous.n`	Compute required sample size for specified confidence level of a prediction interval.
Nonparametric> Plot	`plot.pred.int.npar.design`	Create plots for sampling design based on a prediction interval.

Table 2.2. Sample size menu items and functions for prediction intervals.

For the normal distribution, you can compute the half-width of the prediction interval given the user-specified sample size, compute the required sample size given the user-specified half-width, and plot the relationship between sample size and half-width. For a nonparametric prediction interval, you can compute the required sample size for a specified confidence level, compute the confidence level associated with a given sample size, and plot the relationship between sample size and confidence level.

2.9.1   Prediction Interval for a Normal Distribution

The help files for the functions `pred.int.norm.half.width` and `pred.int.norm.n` explain how to compute a prediction interval for a normal distribution, as well as the relationship between the sample size, number of future observations the prediction interval should contain, estimated standard deviation, confidence level, and half-width.   The function `pred.int.norm.half.width` computes the half-width associated with the prediction interval, given the sample size, number of future observations, estimated standard deviation, and confidence level.   The function `pred.int.norm.n` computes the sample size required to achieve a specified half-width, given the number of future observations, estimated standard deviation, and confidence level. The function `plot.pred.int.norm.design` plots the relationships between sample size, number of future observations, half-width, estimated standard deviation, and confidence level.

The data frame `arsenic3.df` contains arsenic concentrations (ppb) collected quarterly at a background and compliance well. The estimated mean and standard deviation for the background well are 28 and 17 ppb, respectively, based on a sample size of $n = 12$ quarterly samples. The exact two-sided 95% prediction limit for the next $k = 4$ future observations is [-25, 80], which has a half-width of 52.5 ppb and includes values less than 0, which are not possible to observe. In fact, given an assumed standard deviation of $s = 17$, the smallest half width you can achieve for a prediction interval for the next $k = 4$ future observations is 42 ppb, based on an infinite sample size. ***Unlike a confidence interval, the half-width of a prediction interval does not approach 0 as the sample size increases.*** Figure 2.4 shows a plot of sample size versus half-width for a 95% prediction interval for a normal distribution for various values of k (the number of future observations), assuming a standard deviation of $s = 17$.

Menu

To create the plot shown in Figure 2.4, follow these steps.

1.   On the S-PLUS menu bar, make the following menu choices: **EnvironmentalStats>Sample Size and Power>Prediction Intervals>Normal>Plot**. This will bring up the Plot Normal PI Design dialog box.
2.   For X Variable select **Sample Size**, for Y Variable select **Half-Width**, for Minimum X type **4**, for Maximum X type **50**, for # Future Obs type **4**, for PI Method select **Bonferonni**, for Confidence Level (%) select **95**, and for Standard Deviation type **17**.
3.   Click on the **Plotting** tab. **Uncheck** the Add Title box. In the X-Axis Limits box type **0, 50**. In the Y-Axis Limits box type **30, 110**. Click **Apply**.

4. Click on the **Model** tab. For # Future Obs type **2**. Click on the **Plotting** tab, **check** the Add to Current Plot box, and for Line Type select the fourth pattern. Click **Apply**.

5. Click on the **Model** tab. For # Future Obs type **1**. Click on the **Plotting** tab, and for Line Type select the eighth pattern. Click **OK**.

Command

To create the plot shown in Figure 2.4, type these commands.

```
> plot.pred.int.norm.design(range.x.var=c(4, 50), k=4,
    sigma.hat=17, xlim=c(0, 50), ylim=c(30, 110),
    main="Half-Width vs. Sample Size for Prediction
    Interval\nWith Sigma.Hat=17 and Various Values of k")
> plot.pred.int.norm.design(range.x.var=c(4, 50), k=2,
    sigma.hat=17, plot.lty=4, add=T)
> plot.pred.int.norm.design(range.x.var=c(4, 50), k=1,
    sigma.hat=17, plot.lty=2, add=T)
> legend(25, 100, c("k=4", "k=2", "k=1"), lty=c(1, 4, 2),
    lwd=1.5)
```

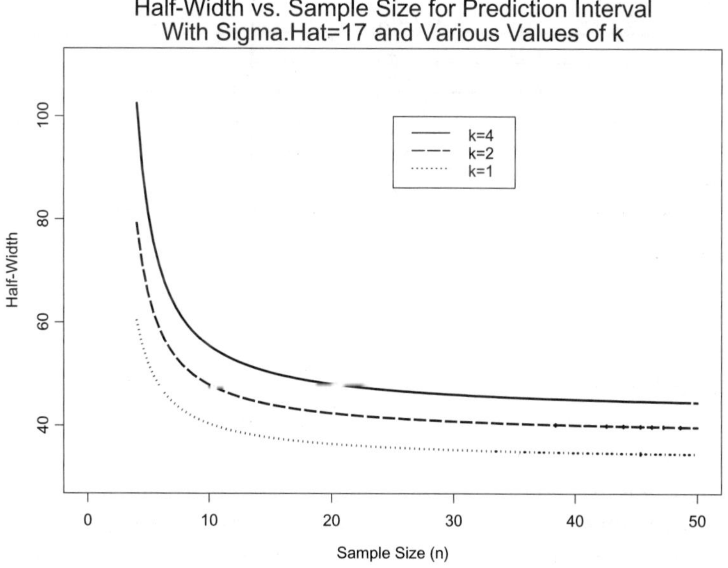

Figure 2.4. The half-width of a prediction interval for arsenic concentrations (ppb) as a function of sample size and number of future observations (k).

2.9.2   Nonparametric Prediction Interval

The help files for the functions `pred.int.npar.conf.level` and `pred.int.npar.n` explain how to compute a nonparametric prediction interval, as well as the relationship between the sample size, the minimum number of future observations the interval should contain (k), the number of future observations (m), and the confidence level. The function `pred.int.npar.conf.level` computes the confidence level associated with the prediction interval, given the number of future observations and sample size. The function `pred.int.npar.n` computes the sample size required to achieve a specified confidence level, given the number of future observations. The function `plot.pred.int.npar.design` plots the relationships between sample size, confidence level, and number of future observations.

Table 2.3 shows the required sample size for a two-sided nonparametric prediction interval for the next m future observations (assuming $k = m$) for various values of m and required confidence levels, assuming we are using the minimum and maximum values as the prediction limits. Figure 2.5 displays the confidence level of a two-sided nonparametric prediction interval as a function of sample size for various values of m.

Confidence Level (%)	# Future Observations (m)	Required Sample Size (n)
90	1	19
	5	93
	10	186
95	1	39
	5	193
	10	386

Table 2.3. Required sample sizes for a two-sided nonparametric prediction interval.

Menu

To compute the required sample sizes for the specified confidence levels and number of future observations using the ENVIRONMENTALSTATS for S-PLUS pull-down menu, follow these steps.

1. On the S-PLUS menu bar, make the following menu choices: **EnvironmentalStats>Sample Size and Power>Prediction Intervals>Nonparametric>Compute**. This will bring up the Nonparametric PI Confidence Level and Sample Size dialog box.

2. For Compute select **Sample Size, uncheck** the Simultaneous box, for # Future Obs and Min # Obs PI Should Contain type **1**, for CI Type select **two-sided**, and for Confidence Levels (%) select **90** and **95**. Click **Apply**.

3. Repeat Step 2, except for # Future Obs and Min # Obs PI Should Contain type **5**. Repeat again but type **10** instead.

To create the plot shown in Figure 2.5, follow these steps.

1. On the S-PLUS menu bar, make the following menu choices: **EnvironmentalStats>Sample Size and Power>Prediction Intervals>Nonparametric>Plot**. This will bring up the Plot Nonparametric PI Design dialog box.
2. For X Variable select **Sample Size**, for Y Variable select **Confidence**, for Minimum X type **2**, for Maximum X type **100**, **uncheck** the Simultaneous box, for # Future Obs and Min # Future Obs PI Should Contain type **1**, and for PI Type select **two-sided**.
3. Click on the **Plotting** tab. **Uncheck** the Add Title box. In the X-Axis Limits box type **0, 100**. In the Y-Axis Limits box type **0, 1**. Click **Apply**.
4. Click on the **Model** tab. For # Future Obs type and Min # Future Obs PI Should Contain type **5**. Click on the **Plotting** tab, **check** the Add to Current Plot box, and for Line Type select the fourth pattern. Click **Apply**.
5. Click on the **Model** tab. For # Future Obs type and Min # Future Obs PI Should Contain type **10**. Click on the **Plotting** tab, and for Line Type select the eighth pattern. Click **OK**.

Command

To compute the required sample sizes for the specified confidence levels and number of future observations, type this command.

```
> pred.int.npar.n(m=rep(c(1, 5, 10), 2),
    conf.level=rep(c(0.9, 0.95), each=3))
```

To create the plot shown in Figure 2.5, type these commands.

```
> plot.pred.int.npar.design(range.x.var=c(2, 100), k=1,
    m=1, xlim=c(0, 100), ylim=c(0, 1), main="Confidence
    Level vs. Sample Size for Nonparametric Prediction
    Interval\nWith Various Values of m (Two-Sided PI)")
> plot.pred.int.npar.design(range.x.var=c(2, 100), k=5,
    m=5, plot.lty=4, add=T)
> plot.pred.int.npar.design(range.x.var=c(2, 100), k=10,
    m=10, plot.lty=2, add=T)
> legend(60, 0.5, c("m=  1", "m=  5", "m=10"),
    lty=c(1, 4, 2), lwd=1.5)
```

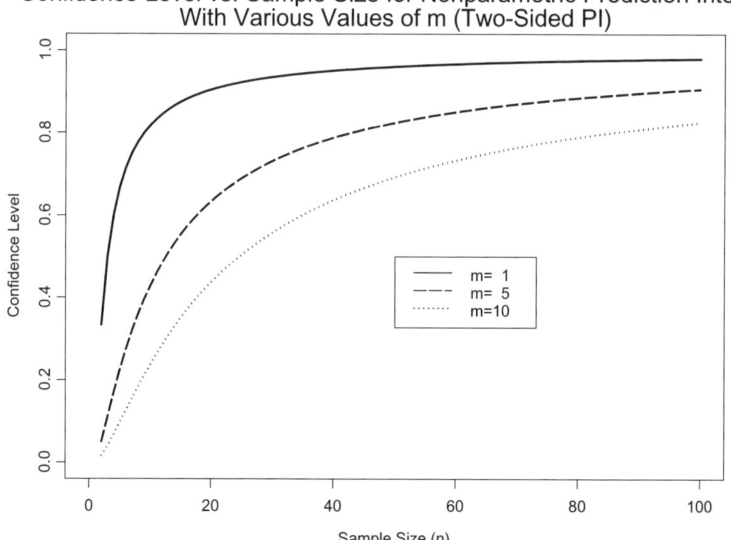

Figure 2.5. The confidence level of a two-sided nonparametric prediction interval as a function of sample size, for various values of the number of future observations (m).

2.10 Sample Size Calculations for Tolerance Intervals

Table 2.4 lists the menu items and functions available in ENVIRONMENTALSTATS for S-PLUS for computing required sample sizes, half-widths, coverage, and confidence levels associated with a tolerance interval. All menu items fall under the menu choice **EnvironmentalStats>Sample Size and Power>Tolerance Intervals**. For example, the first entry in the table describes what you can do if you make the selection **EnvironmentalStats>Sample Size and Power>Tolerance Intervals>Normal>Compute**.

For the normal distribution, you can compute the half-width of the tolerance interval given the user-specified sample size and coverage, compute the required sample size given the user-specified half-width and coverage, and plot the relationship between sample size, half-width, and coverage. For a nonparametric prediction interval, you can compute the required sample size for a specified confidence level and coverage, compute the confidence level associated with a given sample size and coverage, compute the coverage associated with a given sample size and confidence level, and plot the relationship between sample size, confidence level, and coverage.

Menu Item	Functions	Description
Normal> Compute	`tol.int.norm.half.width`	Compute half-width of a tolerance interval for a normal distribution.
	`tol.int.norm.n`	Compute required sample size for specified half-width of a tolerance interval for a normal distribution.
Normal> Plot	`plot.tol.int.norm.design`	Create plots for sampling design based on tolerance interval for a normal distribution.
Nonparametric> Compute	`tol.int.npar.conf.level`	Compute confidence level of a tolerance interval, given the coverage and sample size.
	`tol.int.npar.coverage`	Compute coverage of a tolerance interval, given the confidence level and sample size
	`tol.int.npar.n`	Compute required sample size for specified confidence level and coverage of a tolerance interval.
Nonparametric> Plot	`plot.tol.int.npar.design`	Create plots for sampling design based on a tolerance interval.

Table 2.4. Sample size menu items and functions for tolerance intervals.

2.10.1 Tolerance Interval for a Normal Distribution

The help files for the functions `tol.int.norm.half.width` and `tol.int.norm.n` explain how to compute a tolerance interval for a normal distribution, as well as the relationship between the sample size, estimated standard deviation, confidence level, coverage, and half-width. The function `tol.int.norm.half.width` computes the half-width associated with the tolerance interval, given the sample size, coverage, estimated standard deviation, and confidence level. The function `tol.int.norm.n` computes the sample size required to achieve a specified half-width, given the coverage, estimated standard deviation, and confidence level. The function `plot.tol.int.norm.design` plots the relationships between sample size, half-width, coverage, estimated standard deviation, and confidence level.

Again using the data frame `arsenic3.df` containing arsenic concentrations, we saw that the estimated mean and standard deviation for the background well are 28 and 17 ppb, respectively, based on a sample size of $n = 12$ quarterly samples. The exact two-sided β-content tolerance limit with 95% coverage and associated confidence level of 99% is [-39, 94], which has a half-width of 66.5 ppb and includes values less than 0, which are not possible to observe. In fact, given an assumed standard deviation of $s = 17$, the smallest half width you can achieve for a tolerance interval with 95% coverage and 99% confidence is 33 ppb, based on an infinite sample size. *Unlike a confidence interval, the half-width of a tolerance interval does not approach 0 as the sample size increases.*

Figure 2.6 shows a plot of sample size versus half-width for a β-content tolerance interval for a normal distribution with confidence level 99% for various values of coverage, assuming a standard deviation of $s = 17$.

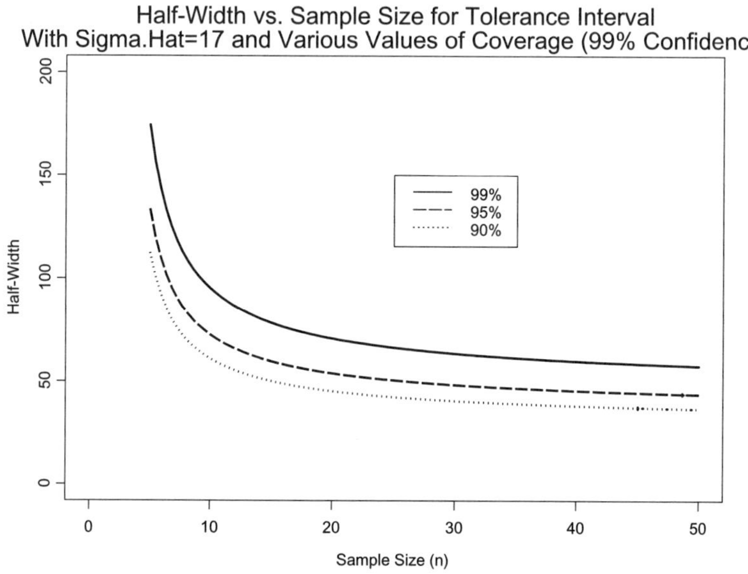

Figure 2.6. The half-width of a tolerance interval for arsenic concentrations (ppb) as a function of sample size and coverage.

Menu

To create the plot shown in Figure 2.6, follow these steps.

1. On the S-PLUS menu bar, make the following menu choices: **EnvironmentalStats>Sample Size and Power>Tolerance Intervals>Normal>Plot**. This will bring up the Plot Normal TI Design dialog box.
2. For X Variable select **Sample Size**, for Y Variable select **Half-Width**, for Minimum X type **5**, for Maximum X type **50**, for Coverage Type select **content**, for Confidence Level (%) select **99**, for Coverage(%) select **99**, and for Standard Deviation type **17**.
3. Click on the **Plotting** tab. **Uncheck** the Add Title box. In the X-Axis Limits box type **0, 50**. In the Y-Axis Limits box type **0, 200**. Click **Apply**.
4. Click on the **Model** tab. For Coverage(%) select **95**. Click on the **Plotting** tab, **check** the Add to Current Plot box, and for Line Type select the fourth pattern. Click **Apply**.
5. Click on the **Model** tab. For Coverage (%) select **90**. Click on the **Plotting** tab, and for Line Type select the eighth pattern. Click **OK**.

Command

To create the plot shown in Figure 2.6, type these commands.

```
> plot.tol.int.norm.design(range.x.var=c(5, 50),
    sigma.hat=17, coverage=0.99, conf=0.99,
    xlim=c(0, 50), ylim=c(0, 200), main="Half-Width vs.
    Sample Size for Tolerance Interval\nWith Sigma.Hat=17
    and Various Values of Coverage (99% Confidence)")
> plot.tol.int.norm.design(range.x.var=c(5, 50),
    sigma.hat=17, coverage=0.95, conf=0.99, plot.lty=4,
    add=T)
> plot.tol.int.norm.design(range.x.var=c(5, 50),
    sigma.hat=17, coverage=0.90, conf=0.99, plot.lty=2,
    add=T)
> legend(25, 150, c("99%", "95%", "90%"), lty=c(1, 4, 2),
    lwd=1.5)
```

2.10.2 Nonparametric Tolerance Interval

The help files for the functions `tol.int.npar.conf.level`, `tol.int.npar.coverage`, and `tol.int.npar.n` explain how to compute a nonparametric tolerance interval, as well as the relationship between the sample size, the coverage, and the confidence level. The function `tol.int.npar.conf.level` computes the confidence level associated with the tolerance interval, given the coverage and sample size. The function `tol.int.npar.coverage` computes the coverage associated with the tolerance interval, given the confidence level and sample size. The function `tol.int.npar.n` computes the sample size required to achieve a specified confidence level, for a given coverage. The function `plot.tol.int.npar.design` plots the relationships between sample size, confidence level, and coverage.

Confidence Level (%)	Coverage (%)	Required Sample Size (n)
90	80	18
	90	38
	95	77
95	80	22
	90	46
	95	93

Table 2.5. Required sample sizes for a two-sided nonparametric tolerance interval.

Table 2.5 shows the required sample size for a two-sided nonparametric tolerance interval for various values of coverage and required confidence levels, assuming we are using the minimum and maximum values as the tolerance limits. Figure 2.7 displays the confidence level of a two-sided nonparametric tolerance interval as a function of sample size for various values of coverage.

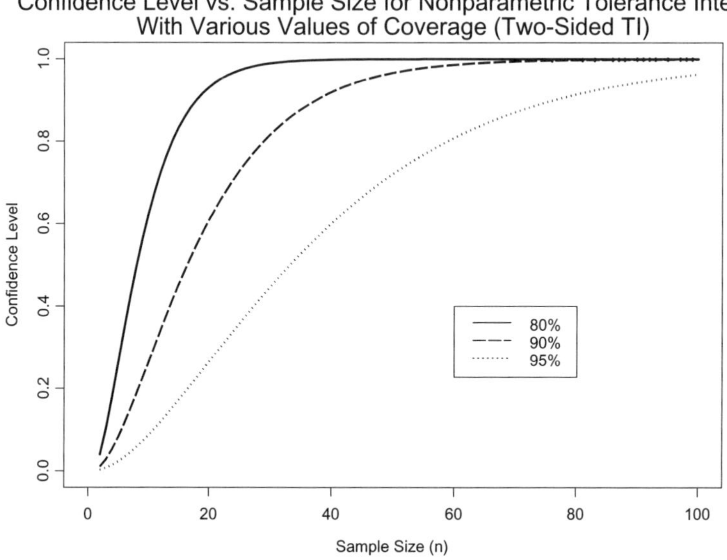

Figure 2.7. The confidence level of a two-sided nonparametric tolerance interval as a function of sample size, for various values of coverage.

Menu

To compute the required sample sizes for the specified confidence levels and coverage using the ENVIRONMENTALSTATS for S-PLUS pull-down menu, follow these steps.

1. On the S-PLUS menu bar, make the following menu choices: **EnvironmentalStats>Sample Size and Power>Tolerance Intervals>Nonparametric>Compute**. This will bring up the Nonparametric TI Confidence Level and Sample Size dialog box.
2. For Compute select **Sample Size**, for TI Type select **two-sided**, for Coverage Type select **content**, for Confidence Level(s) (%) select **90**, and for Coverage(s)(%) select **80**, **90**, and **95**. Click **Apply**.
3. Repeat Step 2, except for Confidence Level(s)(%) select **95**.

To create the plot shown in Figure 2.7, follow these steps.

1. On the S-PLUS menu bar, make the following menu choices:
    EnvironmentalStats>Sample Size and Power>Tolerance Intervals>Nonparametric>Plot. This will bring up the Plot Nonparametric
    TI Design dialog box.
2. For X Variable select **Sample Size**, for Y Variable select **Confidence**,
    for Minimum X type **2**, for Maximum X type **100**, for TI Type select
    two-sided, and for Coverage(%) select **80**.
3. Click on the **Plotting** tab. **Uncheck** the Add Title box. In the X-Axis
    Limits box type **0, 100**. In the Y-Axis Limits box type **0, 1**. Click **Apply**.
4. Click on the **Model** tab. For Coverage(%) select **90**. Click on the **Plotting** tab, **check** the Add to Current Plot box, and for Line Type select
    the fourth pattern. Click **Apply**.
5. Click on the **Model** tab. For Coverage(%) select **95**. Click on the **Plotting** tab, and for Line Type select the eighth pattern. Click **OK**.

Command

To compute the required sample sizes for the specified confidence levels and
coverage, type this command.

```
> tol.int.npar.n(coverage=rep(c(0.8, 0.9, 0.95), 2),
    conf.level=rep(c(0.9, 0.95), each=3))
```

To create the plot shown in Figure 2.7, type these commands.

```
> plot.tol.int.npar.design(range.x.var=c(2, 100),
    coverage=0.8, xlim=c(0, 100), ylim=c(0, 1),
    main="Confidence Level vs. Sample Size for
    Nonparametric Tolerance Interval\nWith Various Values
    of Coverage (Two-Sided TI)")
> plot.tol.int.npar.design(range.x.var=c(2, 100),
    coverage=0.90, plot.lty=4, add=T)
> plot.tol.int.npar.design(range.x.var=c(2, 100),
    coverage=0.95, plot.lty=2, add=T)
> legend(60, 0.4, c("80%", "90%", "95%"),
    lty=c(1, 4, 2), lwd=1.5)
```

2.11 Sample Size and Power Calculations for Hypothesis Tests

Table 2.6 lists the menu items and functions available in
ENVIRONMENTALSTATS for S-PLUS for computing required sample sizes, powers, and minimal detectable differences associated with several different hypothesis tests. All menu items fall under the menu choice **EnvironmentalStats>Sample Size and Power>Hypothesis Tests**. For example, the first entry

in the table describes what you can do if you make the selection **Environmen-talStats>Sample Size and Power>Hypothesis Tests>Normal Mean>One-and Two-Samples>Compute**.

Menu Item	Functions	Description
Normal Mean> One- and Two-Samples> Compute	`t.test.power`	Compute power of the one- or two-sample t-test.
	`t.test.n`	Compute required sample size for specified power for one- or two-sample t-test.
	`t.test.scaled.mdd`	Compute required scaled minimal detectable difference (δ/σ) for specified power for one- or two-sample t-test.
Normal Mean> One- and Two-Samples> Plot	`plot.t.test.design`	Create plots for sampling design based on one- or two-sample t-test.
Normal Mean> k Samples> Compute	`aov.power`	Compute power of the F-test for a one-way ANOVA.
	`aov.n`	Compute required sample size for specified power of F-test for a one-way ANOVA.
Normal Mean> k Samples> Plot	`plot.aov.design`	Create plots for sampling design based on F-test for one-way ANOVA.
Lognormal Mean> Compute	`t.test.lnorm.alt.power`	Compute power of the one- or two-sample t-test assuming lognormal distribution.
	`t.test.lnorm.alt.n`	Compute required sample size for specified power for one- or two-sample t-test assuming lognormal distribution.
	`t.test.lnorm.alt.ratio.of.means`	Compute required ratio of means for specified power for one- or two-sample t-test assuming lognormal distribution.
Lognormal Mean> Plot	`plot.t.test.lnorm.alt.design`	Create plots for sampling design based on one- or two-sample t-test assuming lognormal distribution.

Table 2.6. Sample size and power menu items and functions for hypothesis tests.

Menu Item	Functions	Description
Binomial Proportion> Compute	`prop.test.power`	Compute power of the one- or two-sample proportion test.
	`prop.test.n`	Compute required sample size for specified power for one- or two-sample proportion test.
	`prop.test.mdd`	Compute required minimal detectable difference for specified power for one- or two-sample proportion test.
Binomial Proportion> Plot	`plot.prop.test.design`	Create plots for sampling design based on one- or two-sample proportion test.
Normal Linear Trend> Compute	`linear.trend.test.power`	Compute power of the test for non-zero slope.
	`linear.trend.test.n`	Compute required sample size for specified power for the test for non-zero slope.
	`linear.trend.test.scaled.mds`	Compute required minimal detectable slope for specified power for the test for non-zero slope.
Normal Linear Trend> Plot	`plot.linear.trend.test.design`	Create plots for sampling design based on test for non-zero slope.

Table 2.6. (continued). Sample size and power menu items and functions for hypothesis tests.

In this section, we will illustrate using menu items and functions to explore the relationship between sample size and power to test the mean of a normal distribution. See Chapter 8 of Millard and Neerchal (2001) for more examples of exploring the relationship between sample size and power for other kinds of hypothesis tests.

Menu Item	Functions	Description
Prediction Intervals> Normal> Compute	`pred.int.norm.test.power`	Compute power of test based on prediction interval for normal distribution.
	`pred.int.norm.simultaneous.test.power`	Compute power of test based on simultaneous prediction interval for normal distribution.
Prediction Intervals> Normal> Plot	`plot.pred.int.norm.test.power.curve`	Create power curve for test based on prediction interval for normal distribution.
	`plot.pred.int.norm.simultaneous.test.power.curve`	Create power curve for test based on simultaneous prediction interval for normal distribution.
Prediction Intervals> Lognormal> Compute	`pred.int.lnorm.alt.test.power`	Compute power of test based on prediction interval for lognormal distribution.
	`pred.int.lnorm.alt.simultaneous.test.power`	Compute power of test based on simultaneous prediction interval for lognormal distribution.
Prediction Intervals> Lognormal> Plot	`plot.pred.int.lnorm.alt.test.power.curve`	Create power curve for test based on prediction interval for lognormal distribution.
	`plot.pred.int.lnorm.alt.simultaneous.test.power.curve`	Create power curve for test based on simultaneous prediction interval for lognormal distribution.

Table 2.6. (continued). Sample size and power menu items and functions for hypothesis tests.

2.11.1 Testing the Mean of a Normal Distribution

The help files for `t.test.power`, `t.test.n`, and `t.test.scaled.mdd` explain how to use Student's t-test to perform a hypothesis test for the mean of a normal distribution or the difference between two means, as well as the relationship between the sample size, scaled minimal detectable difference (scaled

MDD), type I error level (α-level), and power. The function `t.test.power` computes the power associated with the t-test, given the sample size, scaled MDD, and α-level. The function `t.test.n` computes the sample size required to achieve a specified power, given the scaled MDD and α-level. The function `t.test.scaled.mdd` computes the scaled minimal detectable difference associated with user-specified values of power, sample size, and α-level. The function `plot.t.test.design` plots the relationships between sample size, power, scaled MDD, and α-level.

The guidance document *Statistical Analysis of Ground-Water Monitoring Data at RCRA Facilities* (USEPA, 1989b) contains an example on pages 6-3 to 6-5 that uses aldicarb concentration (ppb) at three different compliance wells. There are four observations at each of the three wells. The data in this example are stored in the data frame `epa.89b.aldicarb1.df` (see the help file *Datasets: USEPA (1989b)*).

```
> epa.89b.aldicarb1.df
   Aldicarb Month Well
1     19.9     1    1
2     29.6     2    1
3     18.7     3    1
4     24.2     4    1
5     23.7     1    2
6     21.9     2    2
7     26.9     3    2
8     26.1     4    2
9      5.6     1    3
10     3.3     2    3
11     2.3     3    3
12     6.9     4    3
```

The Maximum Concentration Limit (MCL) has been set to 7 ppb. We want to test the null hypothesis that the mean aldicarb concentration is less than or equal to 7 versus the alternative that it is greater than 7. The observed means and standard deviations are 23.1 and 4.9 for Well 1, 24.6 and 2.3 for Well 2, and 4.5 and 2.1 for Well 3. The one-sample t-tests to compare the observed means against the hypothesized mean of 7 ppb yield one-sided p-values of 0.004, 0.0003, and 0.95, respectively.

Assuming a population standard deviation of about 5 ppb, for $n = 4$ observations we have a power of only 14% to detect an average aldicarb concentration that is 5 ppb above the MCL (i.e., a scaled MDD of 1) if we set the significance level to 1%. Alternatively, we would need a sample size of at least $n = 16$ to detect an average aldicarb concentration that is 5 ppb above the MCL if we set the significance level to 1% and desire a power of 90%. Figure 2.8 plots power as a function of sample size for a significance level of 1%, assuming a scaled minimal detectable difference of 1. Figure 2.9 plots the scaled minimal detectable difference as a function of sample size for a significance level of 1%, assuming a power of 90%.

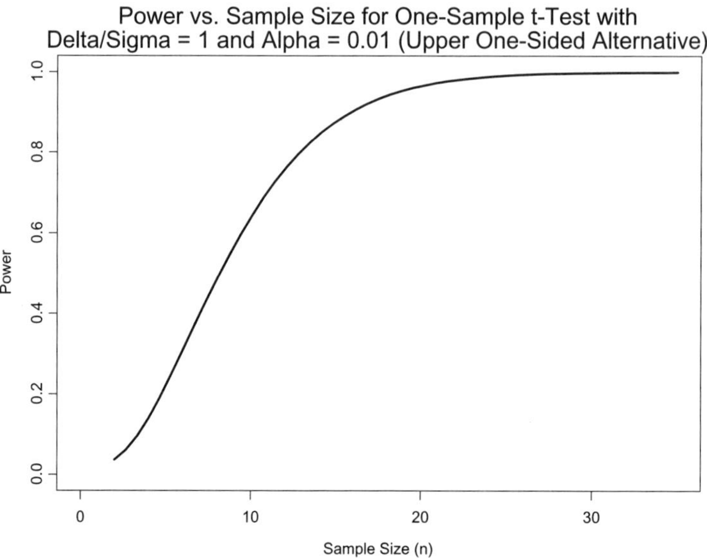

Figure 2.8. Power versus sample size for a significance level of 1%, assuming a scaled minimal detectable difference of $\delta/\sigma = 1$.

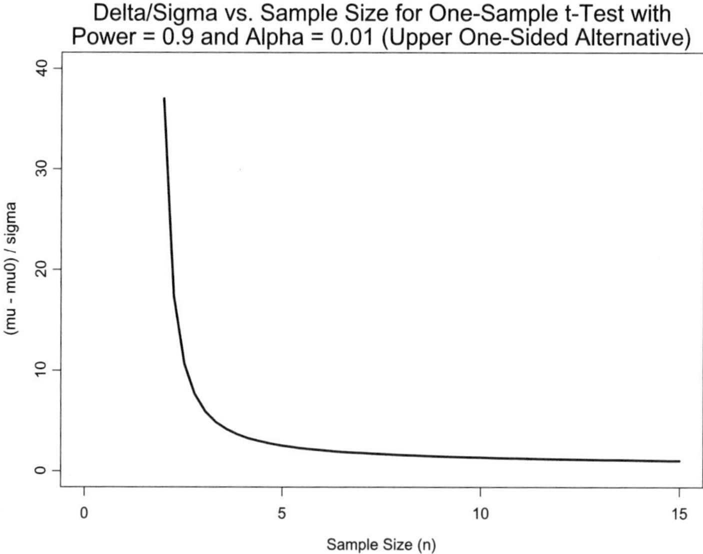

Figure 2.9. Scaled minimal detectable difference versus sample size for a significance level of 1%, assuming a power of 90%.

Menu

To compute the power to detect a scaled minimal detectable difference of 1 assuming a sample size of $n = 4$ and a significance level of 1% using the ENVIRONMENTALSTATS for S-PLUS pull-down menu, follow these steps.

1. On the S-PLUS menu bar, make the following menu choices: **EnvironmentalStats>Sample Size and Power>Hypothesis Tests>Normal Mean>One- and Two-Samples>Compute**. This will bring up the Normal Power and Sample Size dialog box.

2. For Compute select **Power**, for Sample Type select **One-Sample**, for Alpha(s) (%) select **1**, for n type **4**, for Standard Deviation(s) type **5**, for Null Mean type **7**, for Alternative Mean type **12**, and for Alternative select **"greater"**. Click **OK** or **Apply**.

To compute the required sample size to detect a scaled minimal detectable difference of 1 assuming a significance level of 1% and a required power of 90%, repeat Step 1 above, and replace Step 2 with the following:

2. For Compute select **Sample Size**, for Sample Type select **One-Sample**, for Alpha(s) (%) select **1**, for Power(s) (%) select **90**, for Standard Deviation(s) type **5**, for Null Mean type **7**, for Alternative Mean type **12**, and for Alternative select **"greater"**. Click **OK** or **Apply**.

To produce the plot of power versus sample size shown in Figure 2.8 follow these steps.

1. On the S-PLUS menu bar, make the following menu choices: **EnvironmentalStats>Sample Size and Power>Hypothesis Tests>Normal Mean>One- and Two-Samples>Plot**. This will bring up the Plot T-Test Design dialog box.

2. For X Variable select **Sample Size**, for Y Variable select **Power**, for Minimum X type **2**, for Maximum X type **35**, for Sample Type select **One-Sample**, **check** the Use Exact Algorithm box, for Alpha (%) select **1**, for Standard Deviation type **5**, for Null Mean type **7**, for Alternative Mean type **12**, and for Alternative select **"greater"**.

3. Click on the **Plotting** tab. For X-Axis Limits type **0,35**, and for Y-Axis Limits type **0,1**. Click **OK** or **Apply**.

To produce the plot of the scaled minimal detectable difference versus sample size shown in Figure 2.9, repeat Step 1 above and replace Steps 2 and 3 with the following:

2. For X Variable select **Sample Size**, for Y Variable select **Delta/Sigma**, for Minimum X type **2**, for Maximum X type **15**, for Sample Type select **One-Sample**, **check** the Use Exact Algorithm box, for Alpha (%) select **1**, for Power (%) select **90**, and for Alternative select **"greater"**.

3. Click on the **Plotting** tab. For X-Axis Limits type **0,15**, and for Y-Axis Limits type **0,40**. Click **OK** or **Apply**.

Command

To compute the power to detect a scaled minimal detectable difference of 1 assuming a sample size of $n = 4$ and a significance level of 1%, type this command.

```
> t.test.power(n.or.n1=4, delta.over.sigma=1, alpha=0.01,
    alternative="greater", approx=F)
```

To compute the required sample size to detect a scaled minimal detectable difference of 1 assuming a significance level of 1% and a required power of 90%, type this command.

```
> t.test.n(delta.over.sigma=1, alpha=0.01, power=0.9,
    alternative="greater", approx=F)
```

To produce the plot of power versus sample size shown in Figure 2.8 type this command.

```
> plot.t.test.design(alpha=0.01, delta.over.sigma=1,
    range.x.var=c(2, 35), xlim=c(0, 35), ylim=c(0, 1),
    alternative="greater", approx=F)
```

To produce the plot of the scaled minimal detectable difference versus sample size shown in Figure 2.9 type this command.

```
> plot.t.test.design(y.var="delta.over.sigma",
    alpha=0.01, power=0.9, range.x.var=c(2, 15),
    xlim=c(0, 15), ylim=c(0, 40), alternative="greater",
    approx=F)
```

2.12 Summary

- The first and most important step of any environmental study is to design the sampling program.
- ***Probability sampling*** or ***random sampling*** involves using a random mechanism to select samples from the population. All statistical methods used to quantify uncertainty assume some form of random sampling was used to obtain the sample.
- The Data Quality Objectives (DQO) process is a systematic planning tool based on the scientific method. The last two steps involve trading off limits on Type I and Type II errors and sample size.
- You can use the ENVIRONMENTALSTATS for S-PLUS menu items and functions listed in Tables 2.1, 2.2, 2.4, and 2.6 (and the help file *Power and Sample Size Calculations*) to estimate required samples sizes for an environmental study.

3

Looking at Data

3.1 Introduction

Once you have a collection of observations from your environmental study, you should thoroughly examine the data in as many ways as possible and relevant. When the first widely available commercial statistical software packages came out in the 1960s, the emphasis was on statistical summaries of data, such as means, standard deviations, and measures of skew and kurtosis. It is still true that "a picture is worth a thousand words," and no amount of summary or descriptive statistics can replace a good graph to explain your data. John Tukey coined the acronym **EDA**, which stands for ***Exploratory Data Analysis***. Helsel and Hirsch (1992, Chapters 1, 2, and 16), USEPA (1996), and Millard and Neerchal (2001, Chapter 3) give a good overview of statistical and graphical methods for exploring environmental data. Cleveland (1993, 1994) and Chambers et al. (1983) are excellent general references for methods of graphing data. This chapter discusses the menu items and functions available in ENVIRONMENTALSTATS for S-PLUS for producing summary statistics and graphs to describe and look at environmental data.

Menu Item	Functions	Description
Summary Statistics	`full.summary`	Compute summary statistics.
CDF Plot> Empirical CDF	`ecdfplot`	Create an empirical CDF plot.
CDF Plot> Compare Two CDFs	`cdf.compare`	Compare an empirical CDF to a hypothesized CDF, or compare two empirical CDFs.
QQ Plot> QQ Plot	`qqplot`	Create a Q-Q plot comparing data to a theoretical distribution or comparing two data sets.
QQ Plot> QQ Plot Gestalt	`qqplot.gestalt`	Create numerous Q-Q plots based on a specified distribution.
Box-Cox Transformations	`boxcox`	Determine an optimal Box-Cox transformation.

Table 3.1. Menu items and functions in ENVIRONMENTALSTATS for S-PLUS for exploratory data analysis.

3.2   EDA Using ENVIRONMENTALSTATS for S-PLUS

S-PLUS comes with numerous menu items and functions for producing summary statistics and graphs to look at your data. Table 3.1 lists the additional menu items and functions available in ENVIRONMENTALSTATS for S-PLUS for performing EDA. All menu items fall under the menu choice **Environmental-Stats>EDA**. For example, the first entry in the table describes what you can do if you make the selection **EnvironmentalStats>EDA>Summary Statistics**.

Statistic	What It Measures / How It Is Computed	Robust to Extreme Values?
Mean	Center of distribution Sum of observations divided by sample size Where the histogram balances	No
Trimmed Mean	Center of distribution Trim off extreme observations and compute mean Where the trimmed histogram balances	Somewhat, depends on amount of trim
Median	Center of the distribution Middle value or mean of middle values Half of observations are less and half are greater	Very
Geometric Mean	Center of distribution Exponentiated mean of log-transformed observations Estimates true median for a lognormal distribution	Yes
Variance	Spread of distribution Average of squared distances from the mean	No
Standard Deviation	Spread of distribution Square root of variance In same units as original observations	No
Range	Spread of distribution Maximum minus minimum	No
Interquartile Range	Spread of distribution 75th percentile minus 25th percentile Range of middle 50% of data	Yes
Median Absolute Deviation	Spread of distribution $1.4826 \times$ Median of distances from the median	Yes
Geometric Standard Deviation	Spread of distribution Exponentiated standard deviation of log-transformed observations	No
Coefficient of Variation	Spread of distribution/Center of distribution Standard deviation divided by mean Sometimes multiplied by 100 and expressed as a percentage	No
Skew	How the distribution leans (left, right, or centered) Average of cubed distances from the mean	No
Kurtosis	Peakedness of the distribution Average of quartic distances from the mean, then subtract 3	No

Table 3.2. A description of commonly used summary statistics.

3.3 Summary Statistics

Summary statistics (also called ***descriptive statistics***) are numbers that you can use to summarize the information contained in a collection of observations. Summary statistics are also called ***sample statistics*** because they are statistics computed from a sample; they do not describe the whole population.

One way to classify summary or descriptive statistics is by what they measure: location (central tendency), spread (variability), skew (long-tail in one direction), kurtosis (peakedness), etc. Another way to classify summary statistics is by how they behave when unusually extreme observations are present: sensitive versus robust. Table 3.2 summarizes several kinds of descriptive statistics based on these two classification schemes.

In the S-PLUS help system, the help files listed under the Contents topic **Mathematical Operations** and the help files listed under the Contents topic **Robust/Resistant Techniques** list the functions available in S-PLUS for computing summary statistics. In the ENVIRONMENTALSTATS for S-PLUS help system, the topic **Summary Statistics** lists additional and/or modified functions for computing summary statistics.

3.3.1 Summary Statistics for TcCB Concentrations

The guidance document USEPA (1994b, pp. 6.22–6.25) contains measures of 1,2,3,4-Tetrachlorobenzene (TcCB) concentrations (ppb) from soil samples at a "Reference" site and a "Cleanup" area. The Cleanup area was previously contaminated and we are interested in determining whether the cleanup process has brought the level of TcCB back down to what you would find in soil typical of that particular geographic region. In ENVIRONMENTALSTATS for S-PLUS, these data are stored in the data frame `epa.94b.tccb.df`.

```
> epa.94b.tccb.df
      TcCB.orig    TcCB Censored      Area
   1       0.22    0.22        F Reference
   2       0.23    0.23        F Reference
   3       0.26    0.26        F Reference
   4       0.27    0.27        F Reference
   .
  47       1.33    1.33        F Reference
  48      <0.09    0.09        T   Cleanup
   .
 121       6.61    6.61        F   Cleanup
 122      18.40   18.40        F   Cleanup
 123      51.97   51.97        F   Cleanup
 124     168.64  168.64        F   Cleanup
```

There are 47 observations from the Reference site and 77 in the Cleanup area. There is one observation in the Cleanup area that was coded as "ND," which stands for nondetect. This means that the concentration of TcCB for this soil sample (if any was present at all) was so small that the procedure used to quantify TcCB concentrations could not reliably measure the true concentration. For the purpose of creating the data frame `epa.94b.tccb.df`, we set the (unreported) detection limit to the value of the smallest observation, which is 0.09. The column `TcCB.orig` displays how the data were originally recorded, the column `TcCB` contains the original observations, except that the nondetect value is set to the censoring level 0.09, the column `Censored` indicates whether the observation was censored (i.e., reported as a nondetect), and the column `Area` indicates which area the observation comes from.

The summary statistics for the TcCB data are shown on page 8 in Chapter 1. The steps to produce these summary statistics from the ENVIRONMENTALSTATS for S-PLUS Menu are shown on page 9, as are the commands to produce the summary statistics from the Command or Script Window. The summary statistics indicate that the observations for the Cleanup area are extremely skewed to the right. The medians for the two areas are about the same, but the mean for the Cleanup area is much larger, indicating a few or more "outlying" observations with large values. This may be indicative of residual contamination that was missed during the cleanup process. Figures 1.1 and 1.2 on page 10 display the histograms and boxplots for the TcCB data.

3.4   Quantile or Empirical CDF Plots

Loosely speaking, the pth *quantile* of a population is the (a) number such that a fraction p of the population is less than or equal to this number. The pth quantile is the same as the $100p^{th}$ percentile; for example, the 0.5 quantile is the same as the 50th percentile. For a population, a plot of the quantiles on the x-axis versus the percentage or fraction of the population less than or equal to that number on the y-axis is called a *cumulative distribution function plot* or *cdf plot* (we will talk more about cumulative distribution functions in Chapter 4). The y-axis is usually labeled as the *cumulative probability* or *cumulative frequency*.

When we have a sample of data from some population, we usually do not know what percentiles our observations correspond to because we do not know the true population percentiles, so we use the sample data to estimate them. A *quantile plot* (also called an *empirical cumulative distribution function plot* or *empirical cdf plot*) plots the ordered data (sorted from smallest to largest) on the x-axis versus the estimated cumulative probabilities on the y-axis (Chambers et al., 1983, pp. 11–19; Cleveland, 1993, pp. 17–20; Cleveland, 1994, pp. 136–139; Helsel and Hirsch, 1992, pp. 21–24). (Sometimes the x- and y-axes are reversed.) The specific formulas that are used to estimate the cumulative probabilities (also called the *plotting positions*) are discussed in Chapter 3 of Millard and Neerchal (2001) and the help file for `ecdfplot`.

3.4.1   Empirical CDFs for the TcCB Data

Figure 1.3 on page 13 shows the quantile plot for the Reference area TcCB data. Based on this plot, you can easily pick out the median as about 0.55 ppb and the quartiles as about 0.4 ppb and 0.75 ppb (compare these numbers to the ones listed on page 8). You can also see that the quantile plot quickly rises, then pretty much levels off after about 0.8 ppb, which indicates that the data are skewed to the right (see the histogram for the Reference area data in Figure 1.1 on page 10). Helsel and Hirsch (1992, p. 22) note that quantile plots, unlike histograms, do not require you to figure out how to divide the data into classes, and, unlike boxplots, all of the data are displayed in the graph.

Figure 1.4 on page 13 shows the quantile plot for the Reference area TcCB data with a fitted lognormal distribution. We see that the lognormal distribution appears to fit these data quite well.

Figure 1.5 on page 14 compares the empirical cdf for the Reference area with the empirical cdf for the Cleanup area for the log-transformed TcCB data. As we saw with the histograms and boxplots, the Cleanup area has quite a few extreme values compared to the reference area.

The steps to produce the plots shown in Figures 1.3–1.5 from the ENVIRONMENTALSTATS for S-PLUS Menu are shown on pages 14–15. The commands to produce these plots are shown on page 15.

3.5   Probability Plots or Quantile-Quantile (Q-Q) Plots

A *probability plot* or *quantile-quantile (Q-Q) plot* is a graphical display invented by Wilk and Gnanadesikan (1968) to compare a data set to a particular probability distribution or to compare it to another data set. The idea is that if two population distributions are exactly the same, then they have the same quantiles (percentiles), so a plot of the quantiles for the first distribution versus the quantiles for the second distribution will fall on the 0-1 line (i.e., the straight line $y = x$ with intercept 0 and slope 1). If the two distributions have the same shape and spread but different locations, then the plot of the quantiles will fall on the line $y = a + x$ (parallel to the 0-1 line) where a denotes the difference in locations. If the distributions have different locations and differ by a multiplicative constant b, then the plot of the quantiles will fall on the line $y = a + bx$ (D'Agostino, 1986a, p. 25; Helsel and Hirsch, 1986, p. 42). Various kinds of differences between distributions will yield various kinds of deviations from a straight line. In ENVIRONMENTALSTATS for S-PLUS, you can add a fitted regression line, a robust regression line, or a 0-1 line to the Q-Q plot.

Instead of adding a fitted regression line to a Q-Q plot, another way to assess deviation from linearity is to use a Tukey mean-difference Q-Q plot, also called an m-d plot (Cleveland, 1993, pp. 22–23). This is a plot of the differences between the quantiles on the y-axis versus the average of the quantiles on the x-axis. If the two sets of quantiles come from the same parent distribution, then the points in an m-d plot should fall roughly along the horizontal line $y = 0$. If one set of quantiles come from the same distribution with a shift in median, then the points in this plot should fall along a horizontal line above or below the line $y = 0$. If the parent distributions of the quantiles differ in scale, then the points on this plot will fall at an angle.

3.5.1   Q-Q Plots for the Normal and Lognormal Distribution

Figure 3.1 shows the normal Q-Q plot for the Reference area TcCB data, along with a fitted regression line. In this figure you can see that the points do not tend to fall on the line, but rather seem to make a U shape. This indicates that the Reference area data are skewed to the right relative to a symmetrical, bell-shaped normal distribution (see Figure 1.1). Figure 1.6 on page 17 shows the normal Q-Q plot for the log-transformed Reference area data, and Figure 1.7 displays the corresponding Tukey mean-difference Q-Q plot. Here you can see the points do tend to fall on the line, indicating that a lognormal distribution may be a good model for these data. Compare these figures to Figure 1.4.

An interesting feature of normal Q-Q plots is that if a sample of data comes from a normal distribution, and the data are plotted against quantiles of a standard normal distribution (with mean 0 and variance 1), then the intercept of the fitted line estimates the mean of the population, and the slope of the fitted line estimates the standard deviation (Nelson, 1982, p. 113; Cleveland, 1993, p. 31). For the fitted line in Figure 1.6, we can eyeball the intercept at about –0.6 and the slope at about 0.5. The actual mean and standard deviation are –0.62 and 0.47, respectively.

Menu

To create the normal Q-Q plot for the Reference area TcCB data shown in Figure 3.1, follow these steps.

1.   On the S-PLUS menu bar, make the following menu choices: **EnvironmentalStats>EDA>Q-Q Plot>Q-Q Plot**. This will bring up the Q-Q Plot dialog box.
2.   For Data Set select or type **epa.94b.tccb.df**. In the x Variable box, choose **TcCB**. In the Subset Rows with box type **Area=="Reference"**. In the Distribution box select **Normal**.
3.   Click on the **Plotting** tab. **Click** on the Add a Line box to select this option. Click **OK** or **Apply**.

To create the normal Q-Q plot for the log-transformed Reference area data shown in Figure 1.6, follow the same steps as above, except in Step 2, in the Distribution box select **Lognormal**. To create the Tukey mean-difference Q-Q plot shown in Figure 1.7, follow the same steps as above, except in Step 2, in the Distribution box select **Lognormal, check** the Estimate Parameters box, and in Step 3 under Plot Type select **Tukey M-D**.

Command

To create the normal Q-Q plot for the Reference area TcCB data shown in Figure 3.1, type these commands.

```
> attach(epa.94b.tccb.df)
> qqplot(TcCB[Area=="Reference"], add.line=T,
      ylab="Quantiles of TcCB (ppb)",
      main="Normal Q-Q Plot for Reference Area TcCB Data")
```

To create the normal Q-Q plot and Tukey mean-difference Q-Q plot for the log-transformed Reference area data shown in Figures 1.6 and 1.7, type the commands listed on page 10.

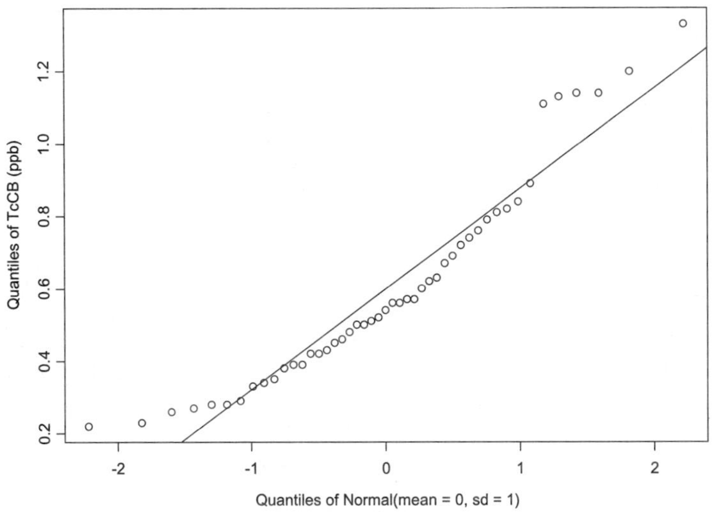

Figure 3.1. Normal Q-Q plot for Reference area TcCB data.

3.5.2 Q-Q Plots for Other Distributions

Although they are not as commonly used as Q-Q plots for the normal and log-normal distributions, you can easily create Q-Q plots for other distributions as well. As an example, the guidance document *Statistical Analysis of Ground-Water Monitoring Data at RCRA Facilities: Addendum to Interim Final Guidance* (USEPA, 1992c, pp.34-40) contains a data set of benzene concentrations (ppb) from water samples collected over six months from six different background monitoring wells. These data are stored in the data frame `epa.92c.benzene1.df` (see the help file *Datasets: USEPA (1992c)*).

```
> epa.92c.benzene1.df
   Benzene.orig Benzene Censored Month Well
 1            <2       2        T     1    1
 2            <2       2        T     2    1
 3            <2       2        T     3    1
 4            <2       2        T     4    1
 .
33           <2       2        T     3    6
34           <2       2        T     4    6
35           10      10        F     5    6
36           <2       2        T     6    6
```

Out of the 36 observations, 33 are reported as "<2", and the other three observations are 10, 12, and 15. The example in the guidance document proposes to model these data as having come from a Poisson distribution, and sets each nondetect to 1 ppb (half the detection limit). Figure 3.2 and Figure 3.3 show the Poisson Q-Q plots for these data, which indicate that the assumption of a Poisson distribution is questionable: there are too many observations with the value 1 (the nondetects), and the detected observations are too large. In Figure 3.2 we indicate multiple observations that have the same (x,y) coordinates with the number of observations that have those coordinates. In Figure 3.3 we instead jitter all of the points.

Menu

To create the Poisson Q-Q plots shown in Figure 3.2 and Figure 3.3, we first need to modify the data frame epa.92c.benzene1.df by adding a column that contains the same data as in the Benzene column, but sets the nondetect values to 1. We will store the result in a new data frame called new.epa.92c.benzene1.df. To do this, follow these steps.

1. Find and highlight **epa.92c.benzene1.df** in the Object Explorer.
2. On the S-PLUS menu bar, make the following menu choices: **Data>Transform**. This will bring up the Transform dialog box.
3. For Data Set, select **epa.92c.benzene1.df**. For Target Column type **New.Benzene**. In the Expression box, type **ifelse(Censored, 1, Benzene)**.
4. Click **OK**. At this point you will get a warning message telling you that you have created a new copy of the data frame epa.92c.benzene1.df that masks the original copy. Close the message window. Also, the modified data frame pops up in a data window. Close this data window.
5. In the left-hand column of the Object Explorer, click on the **Data** folder. In the right-hand column of the Object Explorer, right-click on **epa.92c.benzene1.df** and choose **Properties**. In the Name box rename this data frame to **new.epa.92c.benzene1.df** and click **OK**.

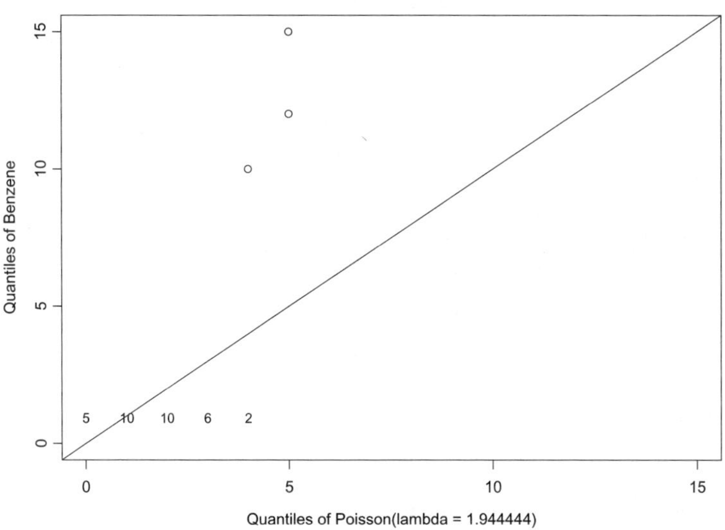

Figure 3.2. Poisson probability plot for the benzene data.

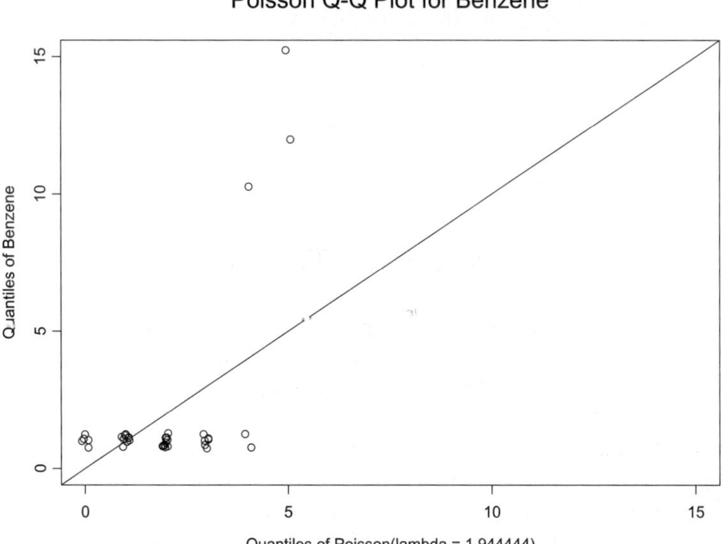

Figure 3.3. Poisson probability plot for the benzene data with jittered points.

To create the Poisson Q-Q plot shown in Figure 3.2, follow these steps:

1. Find and highlight **new.epa.92c.benzene1.df** in the Object Explorer.
2. On the S-PLUS menu bar, make the following menu choices: **EnvironmentalStats>EDA>Q-Q Plot>Q-Q Plot**. This will bring up the Q-Q Plot dialog box.
3. For Data Set select **new.92c.benzene1.df**. In the x Variable box, choose **New.Benzene**. **Check** the Estimate Parameters box. In the Distribution box select **Poisson**.
4. Click on the **Plotting** tab. For Duplicate Points select **Number**. **Check** the Equal Axes box and **check** the Add a Line box to select these options. For Q-Q Line Type select **0-1**. Click **OK** or **Apply**.

To create the Poisson Q-Q plot shown in Figure 3.3, follow the same steps as above, except in Step 4 for Duplicate Points select **Jitter**.

Command

To create the Poisson Q-Q plot shown in Figure 3.2, type these commands.

```
> attach(epa.92c.benzene1.df)
> Benzene[Censored] <- 1
> qqplot(Benzene, dist="pois", estimate.params=T,
    duplicate.points.method="number", add.line=T,
    qq.line.type="0-1")
```

To create the Poisson Q-Q plot shown in Figure 3.3, type these commands.

```
> qqplot(Benzene, dist="pois", estimate.params=T,
    duplicate.points.method="jitter", add.line=T,
    qq.line.type="0-1")
> detach()
```

3.5.3   Using Q-Q Plots to Compare Two Data Sets

Besides using Q-Q plots or probability plots to assess whether a set of data appear to come from a particular probability distribution, you can use a Q-Q plot to assess whether two sets of data appear to have the same parent distribution (i.e., the same shape but not necessarily the same location or scale). If the distributions have the same shape (but not necessarily the same location or scale parameters), then the plot will fall roughly on a straight line. If the distributions are exactly the same, then the plot will fall roughly on the straight line $y = x$.

Figure 3.4 shows the Q-Q plot comparing the Cleanup and Reference areas for the log-transformed TcCB data, along with the 0-1 line. In this figure you can see the points do not tend to fall on the 0-1 line, but instead tend to fall along two different lines, both with a steeper slope than 1. Q-Q plots that exhibit this kind of pattern indicate that one of the samples (the Cleanup area data in this case) probably comes from a "mixture" distribution: some of the observations come from a distribution similar in shape and scale to the Reference area distribution, and some of the observations come from a distribution that is shifted to the right and more spread out relative to the Reference area distribution because of residual contamination.

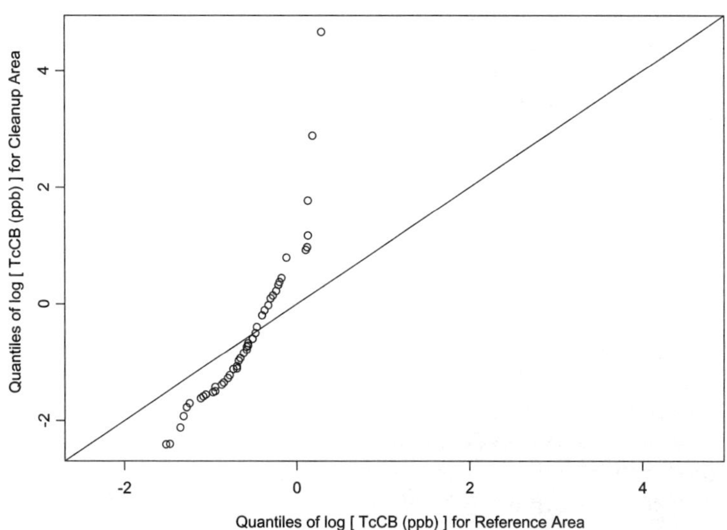

Figure 3.4. Q-Q plot comparing the log-transformed Cleanup and Reference area TcCB data.

Menu

To create the Q-Q plot shown in Figure 3.4, follow these steps.

1. In the Object Explorer, find and highlight the data frame **new.epa.94b.tccb.df** that you created in Section 1.11.3 of Chapter 1.
2. On the S-PLUS menu bar, make the following menu choices: **EnvironmentalStats>EDA>Q-Q Plot>Q-Q Plot**. This will bring up the Q-Q Plot dialog box.
3. Set the Compare Data to button to **Other Data**. The Data Set box should display **new.epa.94b.tccb.df**. In the Variable 1 box, choose **log.TcCB**. In the Variable 2 box, choose **Area**. Click on the **Variable 2 is a Grouping Variable** box to select this option.
4. Click on the **Plotting** tab. **Check** the Equal Axes box. **Check** the Add a Line box. For Q-Q Line Type button, select **0-1**. Click **OK** or **Apply**.

Note that unlike Figure 3.4 this plot has the Reference area quantiles on the *y*-axis and the Cleanup area quantiles on the *x*-axis.

Command

To create the Q-Q plot shown in Figure 3.4, type these commands.

```
> attach(epa.94b.tccb.df)
> qqplot(log(TcCB[Area == "Reference"]),
    log(TcCB[Area == "Cleanup"]), plot.pos.con=0.375,
    equal.axes=T, add.line=T, qq.line.type="0-1",
    xlab=paste("Quantiles of log [ TcCB (ppb) ]",
      "for Reference Area"),
    ylab = paste("Quantiles of log [ TcCB (ppb) ]",
      "for Cleanup Area"),
    main = paste("Q-Q Plot Comparing Cleanup and",
      "Reference Area TcCB Data"))
> detach()
```

3.5.4   Building an Internal Gestalt for Q-Q Plots

Probability plots or Q-Q plots are a graphical, subjective way of assessing the goodness-of-fit of a data set to a specified theoretical distribution. In order to be able to assess the goodness-of-fit, you need to have an internal baseline image (a gestalt) of what a "typical" Q-Q plot looks like when in fact the data come from the specified distribution. You can use ENVIRONMENTALSTATS for S-PLUS to produce numerous Q-Q plots to build up such an internal gestalt (see the help file for qqplot.gestalt for more information).

Figure 3.5 shows a set of four typical normal Q-Q plots based on a sample size of $n = 10$. Note that with such a small number of observations, there can be a bit of spread about the fitted regression line. Figure 3.6 shows a set of four typical Tukey mean-difference Q-Q plots based on a sample size of $n = 10$.

Menu

To create the normal Q-Q plots shown in Figure 3.5, follow these steps.

1.   On the S-PLUS menu bar, make the following menu choices: **EnvironmentalStats>EDA>Q-Q Plot>Q-Q Plot Gestalt**. This will bring up the Q-Q Plot Gestalt dialog box.
2.   **Uncheck** the Estimate Parameters box. For Distribution select **Normal**. For mean type **0** and for sd type **1**. For Sample Size type **10**, for # Pages type **1**, for # Plots/Page type **4**, and for # Rows/Page type **2**.
3.   Click on the **Plotting** tab. **Check** the Add A Line box. Click **OK** or **Apply**.

To create the Tukey mean-difference Q-Q plots shown in Figure 3.6, follow the same steps as above, except in Step 2 **check** the Estimate Parameters box and in Step 3 for Plot Type select **Tukey M-D**. Note that because you cannot set the random seed from the dialog box, your plots will look slightly different.

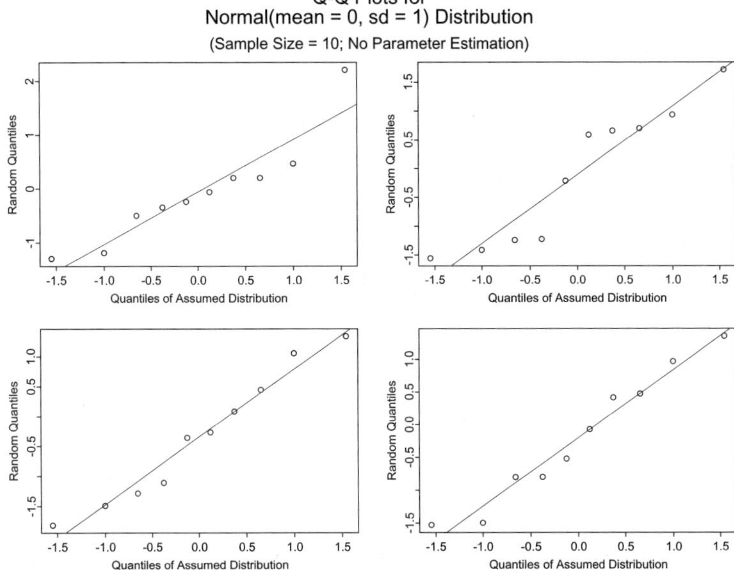

Figure 3.5. Four typical normal Q-Q plots.

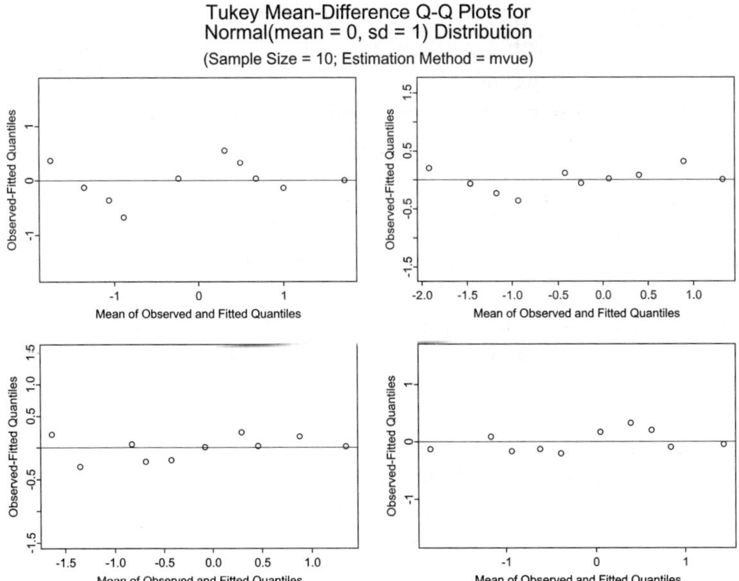

Figure 3.6. Four typical Tukey mean-difference normal Q-Q plots.

Command

To create the normal Q-Q plots shown in Figure 3.5, type these commands.

```
> set.seed(426)
> qqplot.gestalt(num.pages=1, add.line=T)
```

To create the Tukey mean-difference Q-Q plots shown in Figure 3.6, type these commands.

```
> set.seed(426)
> qqplot.gestalt(num.pages=1, add.line=T,
    plot.type="Tukey", estimate.params=T)
```

3.6   Box-Cox Data Transformations and Q-Q Plots

Two common assumptions for several standard parametric hypothesis tests are:

1.   The observations come from a normal distribution.
2.   If several groups are involved, the variances are the same among all of the groups.

A standard linear regression model makes the above assumptions, and also assumes a linear relationship between the response variable and the predictor variable or variables.

Often, especially with environmental data, the above assumptions do not hold because the original data are skewed and/or they follow a distribution that is not really shaped like a normal distribution. It is sometimes possible, however, to transform the original data so that the transformed observations in fact come from a normal distribution or close to a normal distribution. The transformation may also induce homogeneity of variance and a linear relationship between the response and predictor variable(s) (if this is relevant).

Sometimes, theoretical considerations indicate an appropriate transformation. For example, count data often follow a Poisson distribution, and it can be shown that taking the square root of observations from a Poisson distribution tends to make these data look more bell-shaped (Johnson et al., 1992, p. 163; Johnson and Wichern, 1992, p. 164; Zar, 1999, pp. 275–278). A common example in the environmental field is that chemical concentration data often appear to come from a lognormal distribution or some other positively skewed distribution. In this case, taking the logarithm of the observations often appears to yield normally distributed data. Usually, a data transformation is chosen based on knowledge of the process generating the data, as well as graphical tools such as quantile-quantile plots and histograms.

Although data analysts knew about using data transformations for several years, Box and Cox (1964) presented a formalized method for deciding on a data transformation. Given a random variable X from some distribution with only positive values, the Box-Cox family of power transformations is defined as:

$$Y = \begin{cases} \dfrac{\left(X^{\lambda} - 1\right)}{\lambda}, & \lambda \neq 0 \\[2em] \log(X), & \lambda = 1 \end{cases} \tag{3.1}$$

where λ (lambda) denotes the power of the transformation and Y is assumed to come from a normal distribution. This transformation is continuous in λ. Note that this transformation also preserves ordering. That is, if $X_1 < X_2$ then $Y_1 < Y_2$.

Box and Cox (1964) proposed choosing the appropriate value of λ based on maximizing the likelihood function. Note that for non-zero values of λ, instead of using the formula of Box and Cox in Equation (3.1), you may simply use the power transformation

$$Y = X^{\lambda} \tag{3.2}$$

since these two equations differ only by a scale difference and origin shift, and the essential character of the transformed distribution remains unchanged (Draper and Smith, 1981, p. 225).

The value $\lambda = 1$ corresponds to no transformation. Values of λ less than 1 shrink large values of X, and are therefore useful for transforming positively skewed (right-skewed) data. Values of λ larger than 1 inflate large values of X, and are therefore useful for transforming negatively skewed (left-skewed) data (Helsel and Hirsch, 1992, pp. 13–14; Johnson and Wichern, 1992, p. 165). Commonly used values of λ include 0 (log transformation), 0.5 (square-root transformation), -1 (reciprocal), and -0.5 (reciprocal root).

Transformations are not "tricks" used by the data analyst to hide what is going on, but rather useful tools for understanding and dealing with data (Berthouex and Brown, 1994, p. 64). Hoaglin (1988) discusses "hidden" transformations that are used everyday, such as the pH scale for measuring acidity. It is often recommend that when dealing with several similar data sets, it is best to find a common transformation that works reasonably well for all the data sets, rather than using slightly different transformations for each data set (Helsel and Hirsch, 1992, p. 14; Shumway et al., 1989).

One problem with data transformations is that translating results on the transformed scale back to the original scale is not always straightforward. Estimating quantities such as means, variances, and confidence limits in the transformed scale and then transforming them back to the original scale usually leads to biased and inconsistent estimates (Gilbert, 1987, p. 149; Fisher and van Belle, 1993, p. 466). For example, exponentiating the confidence limits for a mean based on log-transformed data does not yield a confidence interval for the mean on the original scale. Instead, this yields a confidence interval for the median. It should be noted, however, that quantiles (percentiles) and rank-based procedures are invariant to monotonic transformations (Helsel and Hirsch, 1992, p. 12).

You can use ENVIRONMENTALSTATS for S-PLUS to determine an "optimal" Box-Cox transformation, based on one of three possible criteria:

- Probability Plot Correlation Coefficient (PPCC)
- Shapiro-Wilk Goodness-of-Fit Test Statistic (W)
- Log-Likelihood Function

You can also compute the value of the selected criterion for a range of values of the transform power λ.

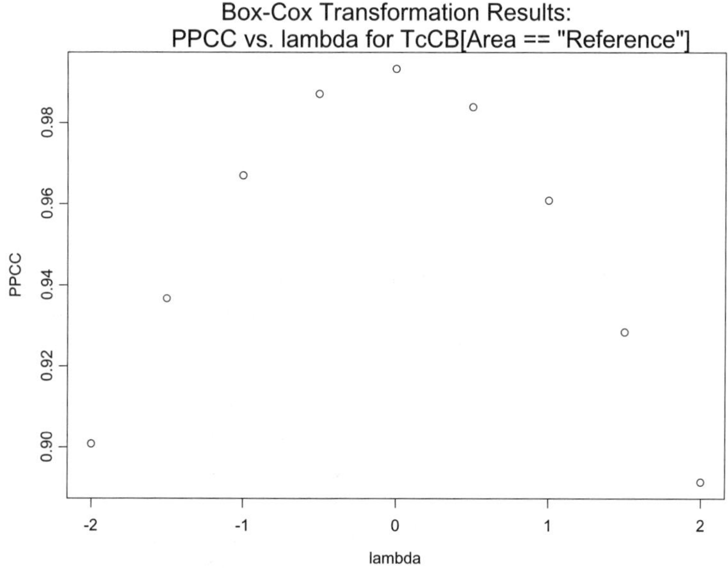

Figure 3.7. Probability plot correlation coefficient versus Box-Cox transform power (λ) for the Reference area TcCB data.

Figure 3.7 displays a plot of the probability plot correlation coefficient versus various values of the transform power λ for the Reference area TcCB data. For this data set, the PPCC reaches its maximum at about $\lambda = 0$, which corresponds to a log transformation. Besides plotting the objective function versus λ, you can also generate Q-Q plots and Tukey mean-difference Q-Q plots for each of the values of λ.

Menu

To create the plot of the PPCC versus the transform power λ for the Reference area TcCB data shown in Figure 3.7, follow these steps.

1.   In the Object Explorer, find and highlight **epa.94b.tccb.df**.

2. On the S-PLUS menu bar, make the following menu choices: **EnvironmentalStats>EDA>Box-Cox Transformations**. This will bring up the Box-Cox Transformations dialog box.

3. For Data to Use select **Pre-Defined Data**. For Data Set select **epa.94b.tccb.df**. In the Variable box, choose **TcCB**. In the Subset Rows with box, type **Area=="Reference"**. Click **OK** or **Apply**.

You will see a print-out of the results showing the value of the PPCC for each of the nine values of the transform power λ. You will also see a graphsheet with 19 pages. The first page is the same as Figure 3.7. The next nine pages display the normal Q-Q plots for each of the values of λ, and the last nine pages display the corresponding Tukey mean-difference Q-Q plots.

Command

To create the plot of the PPCC versus the transform power λ for the Reference area TcCB data shown in Figure 3.7, type these commands.

```
> attach(epa.94b.tccb.df)
> boxcox.list <- boxcox(TcCB[Area=="Reference"])
> plot(boxcox.list)
```

At this point, you will see a menu selection. You can select 2 to plot the PPCC versus λ (Figure 3.7). Selecting 3 will produce the nine Q-Q plots for each of the values of λ, and selecting 4 will produce the nine Tukey mean-difference Q-Q plots. You can create all of these plots by selecting 1. Select 0 to exit the menu.

Alternatively, to create all of the plots, you can type

```
> plot(boxcox.list, plot.type="All")
```

To print the results, type

```
> boxcox.list
```

To detach the data frame `epa.94b.tccb.df` from your search list, type

```
> detach()
```

3.7 Summary

- Summary or descriptive statistics can be classified by what they measure (location, spread, skew, kurtosis, etc.) and also how they behave when unusually extreme observations are present (sensitive versus robust).
- Because environmental data usually involve measures of chemical concentrations, and concentrations cannot fall below 0, environmental data often tend to be positively skewed.

- Graphical displays are usually far superior to summary statistics for conveying information in a data set.
- For conveying the distribution of univariate data, use strip plots, histograms, density plots, boxplots, and quantile plots (also called empirical cumulative distribution function plots).
- To compare two data sets or to compare a data set to a theoretical probability distribution, use quantile-quantile (Q-Q) plots (also called probability plots), and Tukey mean-difference Q-Q plots.
- You can use Box-Cox transformations along with Q-Q plots to determine a transformation that may satisfy the assumption of normality if this assumption is necessary for a hypothesis test or confidence interval.
- You can use the ENVIRONMENTALSTATS for S-PLUS menu items and functions listed in Table 3.1 to compute summary statistics, create empirical cdf plots, create Q-Q plots, and determine "optimal" Box-Cox transformations.

4

Probability Distributions

4.1 Introduction

As we stated in Chapter 2, a population is defined as the entire collection of measurements about which we want to make a statement, such as all possible measurements of dissolved oxygen in a specific section of a stream within a certain time period. Probability distributions are idealized mathematical models that are used to model the variability inherent in a population (e.g., all measures of dissolved oxygen will not exactly match each other). Certain probability distributions come up again and again in environmental statistics. This chapter discusses the menu items and functions available in ENVIRONMENTALSTATS for S-PLUS for plotting probability distributions, computing quantities associated with these distributions, and generating random numbers from these distributions. See Chapter 4 of Millard and Neerchal (2001) for a more in-depth discussion of probability distributions.

Figure 4.1 displays examples of the probability density functions for the probability distributions available in S-PLUS and ENVIRONMENTALSTATS for S-PLUS. Some of these distributions are already available in S-PLUS, and some have been added or modified in ENVIRONMENTALSTATS for S-PLUS. Table 4.1 lists the probability distributions available in S-PLUS and ENVIRONMENTALSTATS for S-PLUS (see the ENVIRONMENTALSTATS for S-PLUS help file *Probability Distributions and Random Numbers*). The help file for `.Distribution.frame` contains a more extensive table that includes the distribution name, abbreviation, type (continuous, discrete, finite discrete, mixed), range (i.e., support), parameters, default values for the parameters, parameter ranges, and estimation methods available for the parameters.

For each of these distributions, there are menu items and functions for computing the probability density function (pdf), the cumulative distribution function (cdf), quantiles, and random numbers. The form of the names of the functions are `dabb`, `pabb`, `qabb`, and `rabb`, where *abb* denotes the abbreviation of the distribution name (see column 2 of Table 4.1). For example, the functions `dnorm`, `pnorm`, `qnorm`, and `rnorm` compute the pdf, cdf, quantiles, and random numbers for the normal distribution.

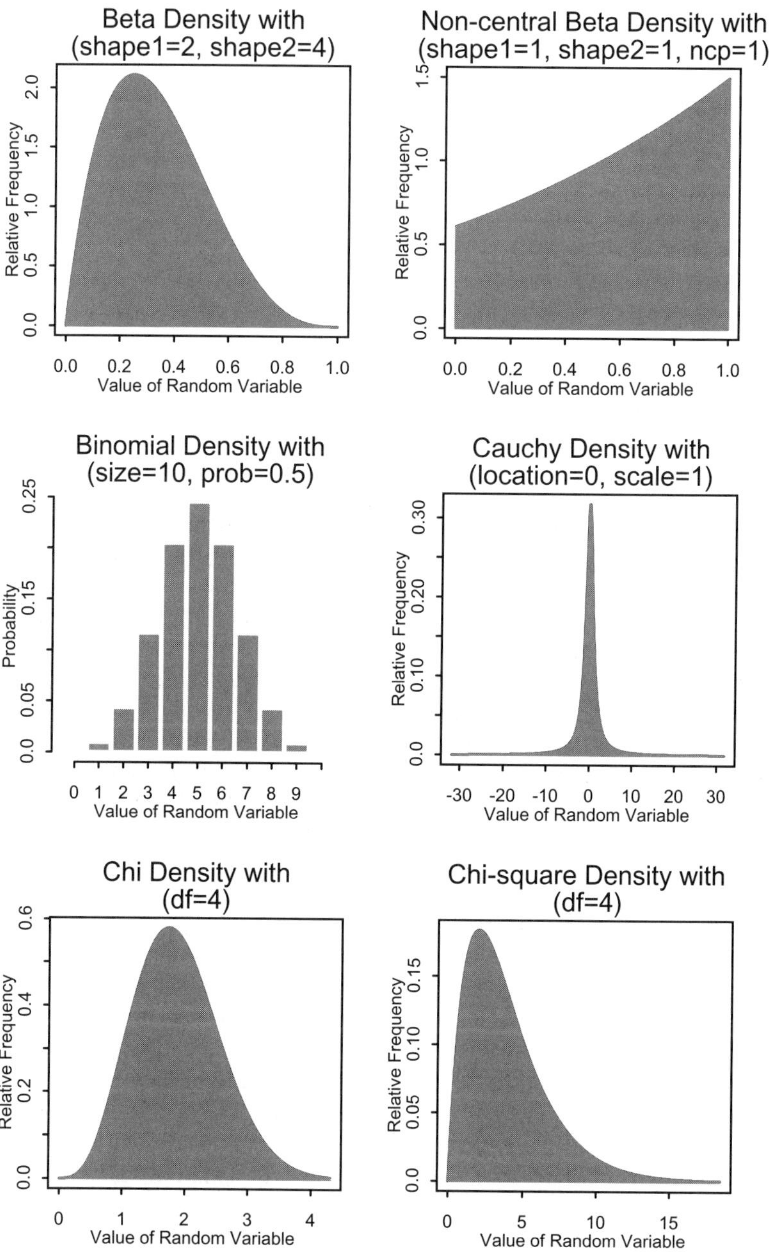

Figure 4.1. Probability distributions in S-PLUS and ENVIRONMENTALSTATS for S-PLUS.

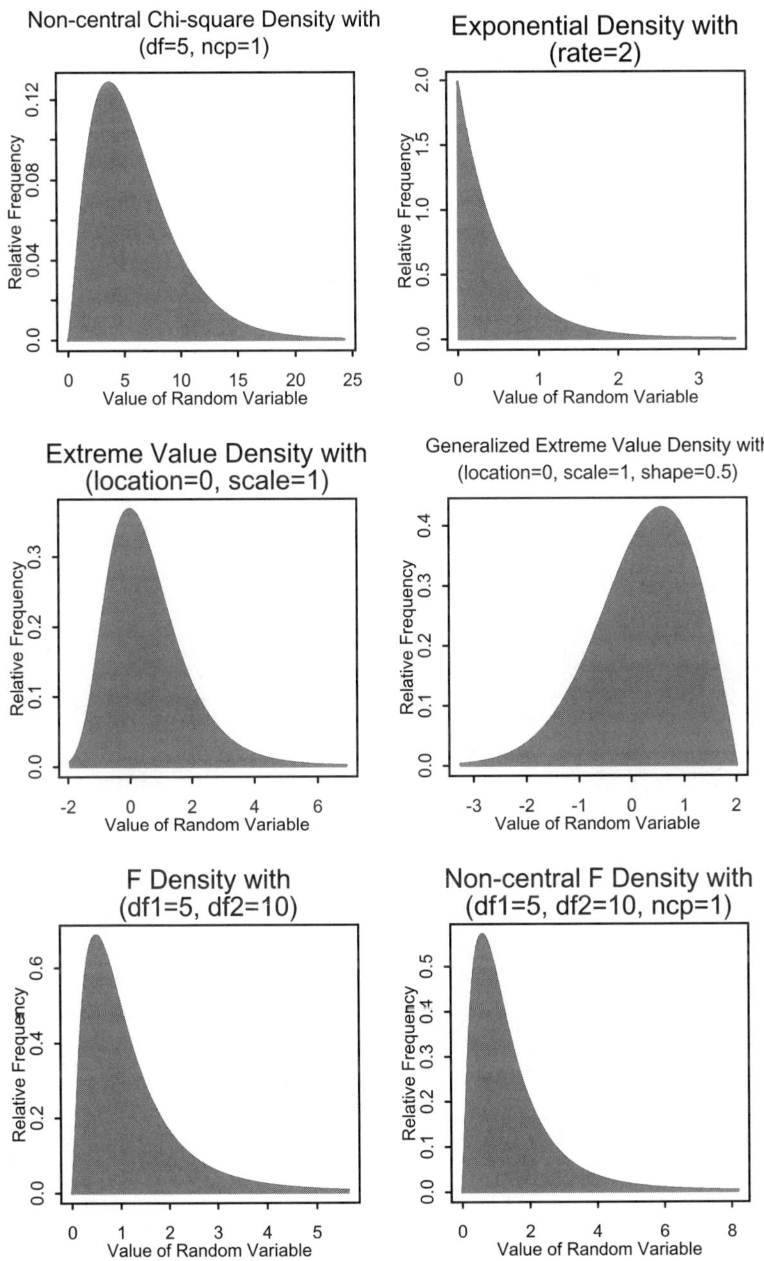

Figure 4.1 (continued). Probability distributions in S-Plus and EnvironmentalStats for S-Plus.

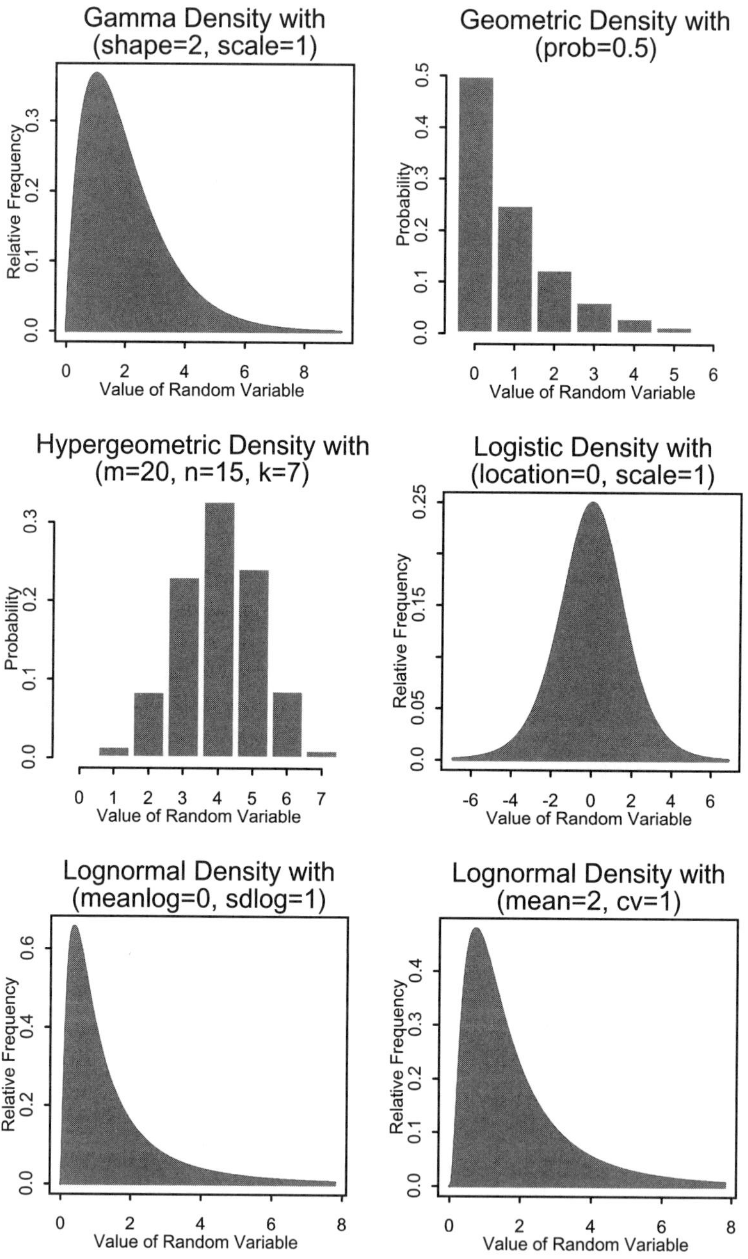

Figure 4.1 (continued). Probability distributions in S-PLUS and ENVIRONMENTALSTATS for S-PLUS.

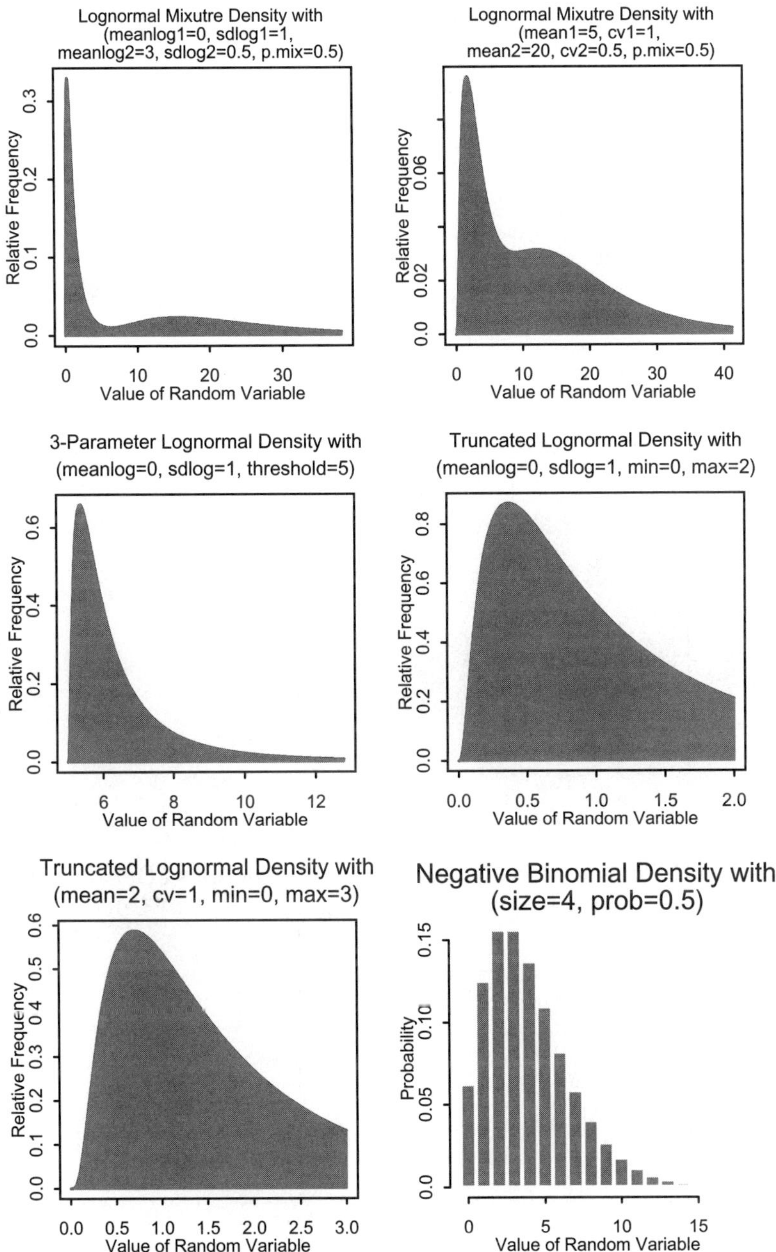

Figure 4.1 (continued). Probability distributions in S-PLUS and ENVIRONMENTALSTATS for S-PLUS.

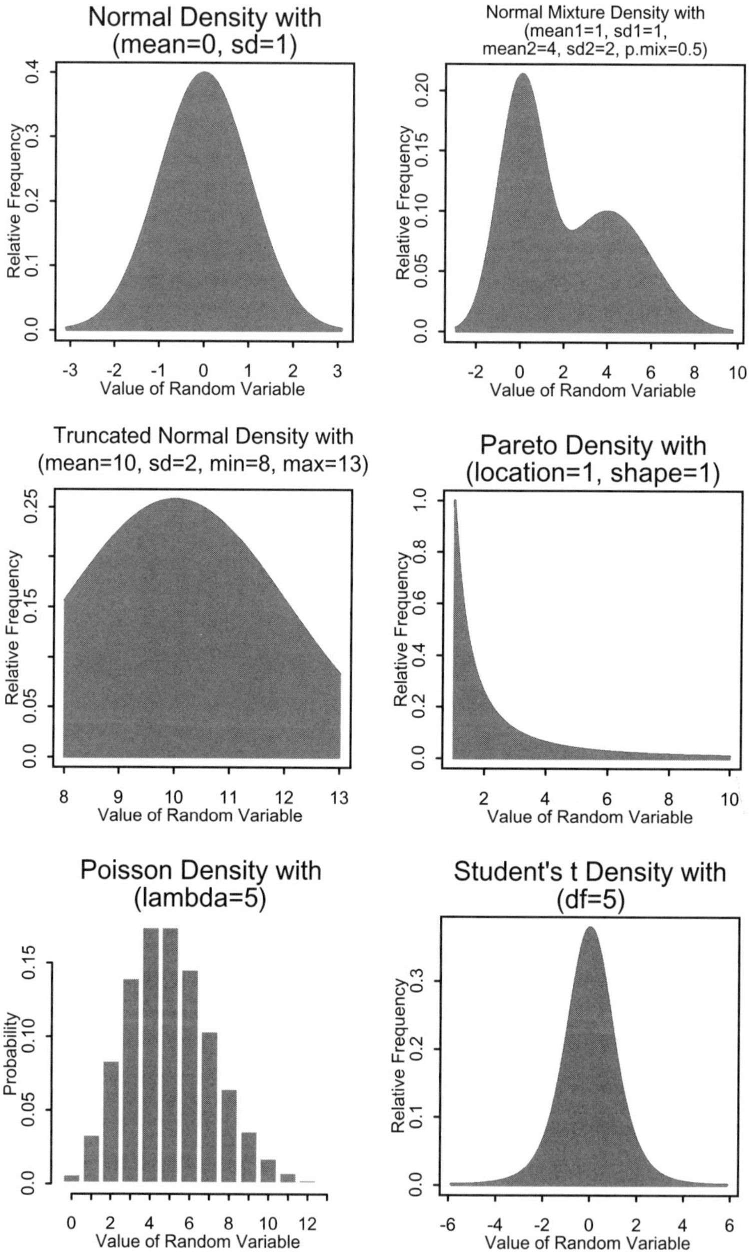

Figure 4.1 (continued). Probability distributions in S-PLUS and ENVIRONMENTALSTATS for S-PLUS.

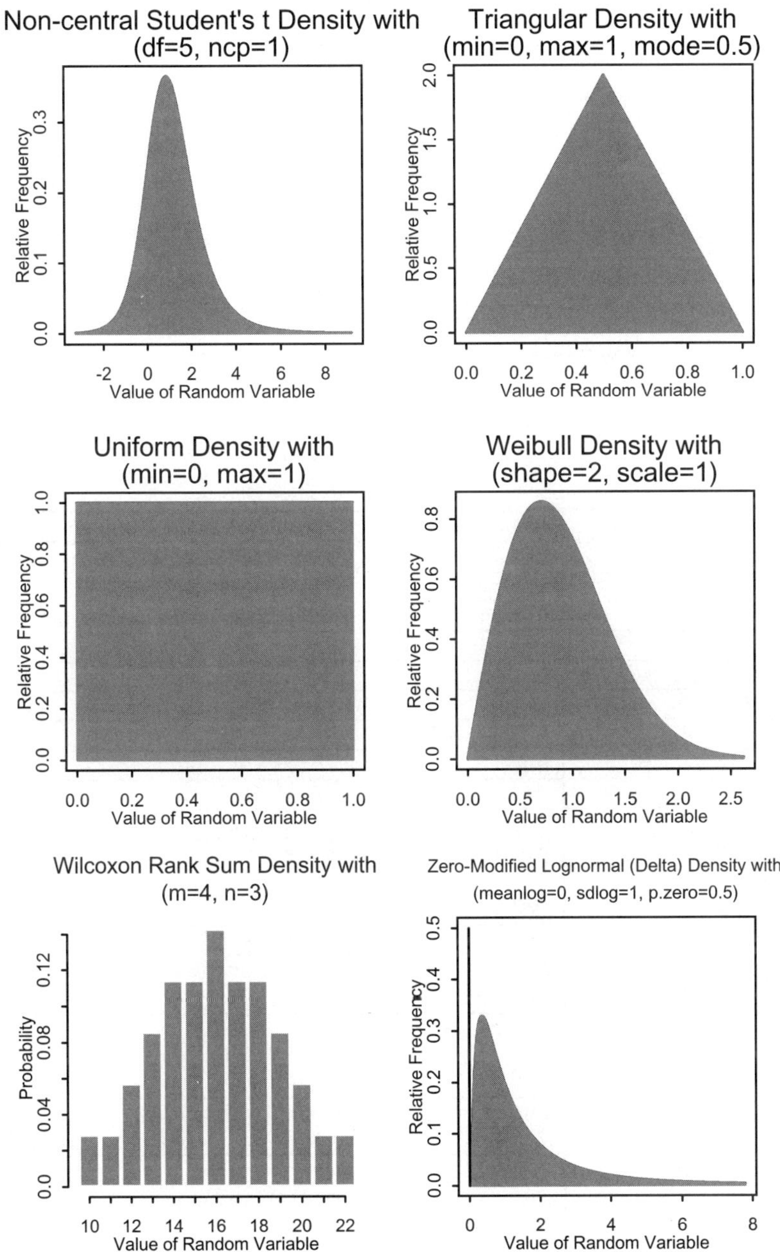

Figure 4.1 (continued). Probability distributions in S-PLUS and ENVIRONMENTALSTATS for S-PLUS.

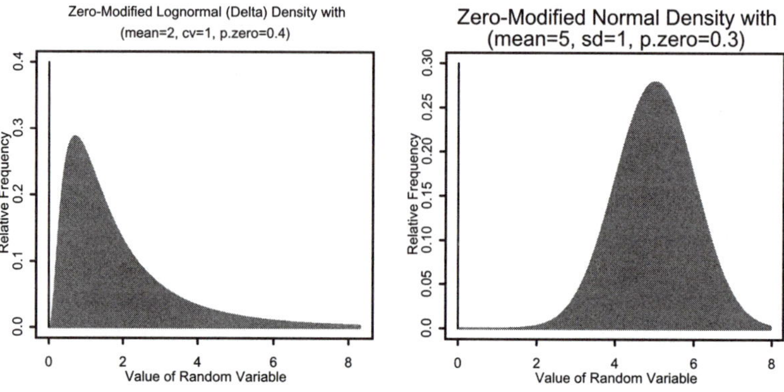

Figure 4.1 (continued). Probability distributions in S-PLUS and ENVIRONMENTALSTATS for S-PLUS.

Distribution Name	Abbreviation	Parameter(s)
Beta*	beta	shape1, shape2, ncp
Binomial	binom	size, prob
Cauchy	cauchy	location, scale
Chi**	chi	df
Chi-square*	chisq	df, ncp
Empirical	emp	
Exponential	exp	rate
Extreme Value**	evd	location, scale
F*	f	df1, df2, ncp
Gamma*	gamma	shape, scale
Generalized Extreme Value**	gevd	location, scale, shape
Geometric	geom.	prob
Hypergeometric	hyper	m, n, k
Logistic	logis	location, scale
Lognormal*	lnorm	meanlog, sdlog
Lognormal (alternative parameterization)**	lnorm.alt	mean, cv
Lognormal Mixture**	lnorm.mix	meanlog1, sdlog1, meanlog2, sdlog2,p.mix
Lognormal Mixture (alternative parameterization)**	lnorm.mix.alt	mean1, sd1, mean2, sd2, cv

Table 4.1. Distribution abbreviations and parameters.

*   Modified in ENVIRONMENTALSTATS for S-PLUS.
**   Available only in ENVIRONMENTALSTATS for S-PLUS.

Distribution Name	Abbreviation	Parameter(s)
3-Parameter Lognormal**	`lnorm3`	`meanlog, sdlog, threshold`
Truncated Lognormal**	`lnorm.trunc`	`meanlog, sdlog, min, max`
Truncated Lognormal (alternative parameterization)**	`lnorm.alt.trunc`	`mean, cv, min, max`
Negative Binomial	`nbinom`	`size, prob`
Normal	`norm`	`mean, sd`
Normal Mixture**	`norm.mix`	`mean1, sd1, mean2, sd2, p.mix`
Truncated Normal**	`norm.trunc`	`mean, sd, min, max`
Pareto**	`pareto`	`location, shape`
Poisson	`pois`	`lambda`
Stable***	`stab`	`index, skewness`
Student's t*	`t`	`df, ncp`
Triangular**	`tri`	`min, max, mode`
Uniform	`unif`	`min, max`
Weibull	`weibull`	`shape, scale`
Wilcoxon Rank Sum	`Wilcox`	`m, n`
Zero-Modified Lognormal (Delta)**	`zmlnorm`	`meanlog, sdlog, p.zero`
Zero-Modified Lognormal (Delta) (alternative parameterization)**	`zmlnorm.alt`	`mean, cv, p.zero`
Zero-Modified Normal**	`zmnorm`	`mean, sd, p.zero`

Table 4.1 (continued). Distribution abbreviations and parameters.

*	Modified in ENVIRONMENTALSTATS for S-PLUS.
**	Available only in ENVIRONMENTALSTATS for S-PLUS.
***	Only random numbers available for stable distribution.

Table 4.2 lists the menu items and functions available in ENVIRONMENTALSTATS for S-PLUS for plotting probability distributions, computing quantities associated with these distributions, and generating random numbers from these distributions. All menu items fall under the menu choice **EnvironmentalStats>Probability Distributions and Random Numbers**. For example, the first entry in the table describes what you can do if you make the selection **EnvironmentalStats>Probability Distributions and Random Numbers>Density, CDF, Quantiles**.

Menu Item	Functions	Description

Density, CDF, Quantiles	`dabb`
	`pabb`
	`qabb`
		Compute values of probability density, cumulative distribution function, or quantiles for the distribution with abbreviation *abb*.

Random Numbers> Univariate	`rabb` `simulate.vector`	Generate random numbers from the distribution with abbreviation *abb*.
Random Numbers> Multivariate> Normal	`rmvnorm`	Generate random numbers from a multivariate normal distribution.
Random Numbers> Multivariate> Based on Rank Correlation	`simulate.mv.matrix`	Generate random numbers from various distributions based on a user-specified rank correlation matrix.
Plot Distribution	`pdfplot` `cdfplot`	Plot the probability density function or the cumulative distribution function.

Table 4.2. Menu items and functions in ENVIRONMENTALSTATS for S-PLUS for probability distributions and random numbers.

4.2   Probability Density Function (PDF)

A ***probability density function (pdf)*** is a mathematical formula that describes the relative frequency of a random variable. Sometimes the picture of this formula is called the pdf. If a random variable is discrete, its probability density function is sometimes called a ***probability mass function***, since it shows the "mass" of probability at each possible value of the random variable.

4.2.1   Probability Density Function for Lognormal Distribution

Figure 4.2 shows the relative frequency (density) histogram for the Reference area TcCB data. For each class (bar) of this histogram, the proportion of observations falling in that class is equal to the area of the bar; that is, it is the width of the bar times the height of the bar. If we could take many, many more samples and create relative frequency histograms with narrower and narrower classes, we might end up with a picture that looks like Figure 4.3, which shows the probability density function of a lognormal random variable with a mean of 0.6 and a coefficient of variation of 0.5. For a continuous random variable, a ***probability distribution*** can be thought of as what a density (relative frequency) histogram of outcomes would look like if you could keep taking more and more samples and making the histogram bars narrower and narrower.

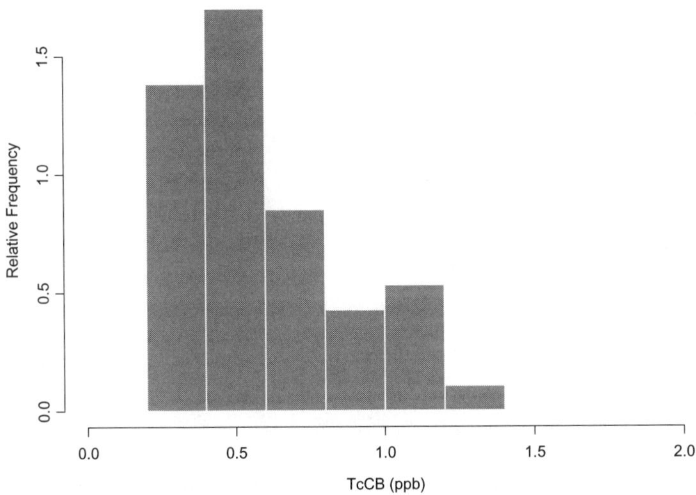

Figure 4.2. Relative frequency (density) histogram of the Reference area TcCB data.

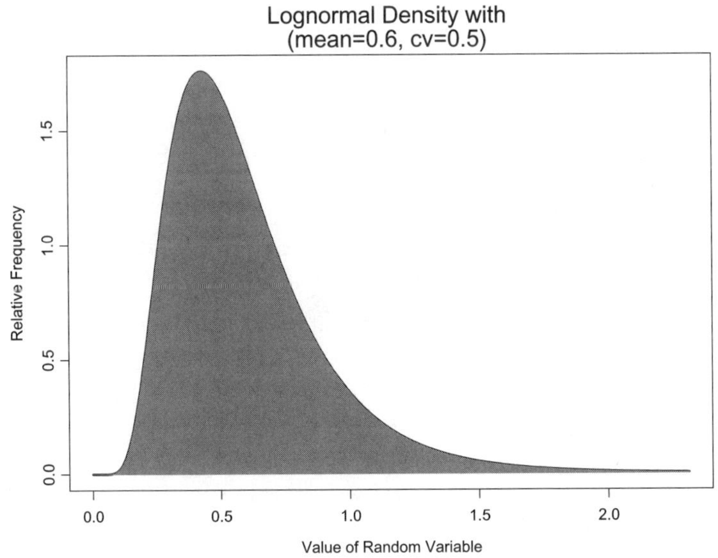

Figure 4.3. A lognormal probability density function.

The probability density function for the lognormal distribution shown Figure 4.3 in is given by:

$$f(x) = \frac{1}{x\sigma\sqrt{2\pi}}\, e^{\frac{-1}{2\sigma^2}\left[\log(x)-\mu\right]^2} \quad,\ x > 0 \tag{4.1}$$

where

$$\mu = \log\left(\frac{\theta}{\sqrt{\tau^2 + 1}}\right)$$

$$\sigma = \sqrt{\log\left(\tau^2 + 1\right)}$$

$$\theta = 0.6 \tag{4.2}$$

$$\tau = 0.5$$

Note that θ and τ denote the mean and coefficient of variation of the distribution, and μ and σ denote the mean and standard deviation of the log-transformed random variable. The values of the pdf evaluated at 0, 0.5, 1, 1.5, and 2 are 0, 1.67, 0.355, 0.053, and 0.009.

Menu

To create the plot of the lognormal pdf shown in Figure 4.3, follow these steps.

1. On the S-PLUS menu bar, make the following menu choices: **EnvironmentalStats>Probability Distributions and Random Numbers>Plot Distribution**. This will bring up the Plot Distribution Function dialog box.
2. In the Distribution box, choose **Lognormal (Alternative)**. In the mean box, type **0.6**. In the cv box, type **0.5**. Click **OK** or **Apply**.

To compute the value of this lognormal pdf at 0, 0.5, 1, 1.5, and 2, follow these steps.

1. On the S-PLUS menu bar, make the following menu choices: **EnvironmentalStats>Probability Distributions and Random Numbers>Density, CDF, Quantiles**. This will bring up the Densities, Cumulative Probabilities, or Quantiles dialog box.
2. For the Data to Use buttons, choose **Expression**. In the Expression box, type **0, 0.5, 1, 1.5, 2**. In the Distribution box, choose **Lognormal (Alternative)**. In the mean box type **0.6**. In the cv box type **0.5**. Under the Probability or Quantile group, **check** the Density box. Click **OK** or **Apply**.

Command

To create the plot of the lognormal pdf shown in Figure 4.3, type this command.

```
> pdfplot(distribution="lnorm.alt",
    param.list=list(mean=0.6, cv=0.5))
```

To compute the value of this lognormal pdf at 0, 0.5, 1, 1.5, and 2, type this command.

```
> dlnorm.alt(c(0, 0.5, 1, 1.5, 2), mean=0.6, cv=0.5)
```

4.3   Cumulative Distribution Function (CDF)

The ***cumulative distribution function (cdf)*** of a random variable X, sometimes called simply the distribution function, is the function F such that

$$F(x) = \Pr(X \le x) \tag{4.3}$$

for all values of x. That is, $F(x)$ is the probability that the random variable X is less than or equal to some number x. The cdf can also be defined or computed in terms of the probability density function (pdf) f as

$$F(x) = \Pr(X \le x) = \int_{-\infty}^{x} f(t)\, dt \tag{4.4}$$

4.3.1   Cumulative Distribution Function for Lognormal Distribution

Figure 4.4 displays the cumulative distribution function for the lognormal random variable whose pdf was shown in Figure 4.3. The values of this cdf at 0, 0.5, 1, 1.5, and 2 are 0, 0.440, 0.906, 0.985, and 0.997.

Menu

To produce the lognormal cdf shown in Figure 4.4, follow these steps.

1. On the S-PLUS menu bar, make the following menu choices: **EnvironmentalStats>Probability Distributions and Random Numbers>Plot Distribution**. This will bring up the Plot Distribution Function dialog box.
2. In the Distribution box, choose **Lognormal (Alternative)**. In the mean box, type **0.6**. In the cv box, type **0.5**. For Plot select **CDF**. Click **OK** or **Apply**.

To compute the value of this lognormal cdf at 0, 0.5, 1, 1.5, and 2, follow these steps.

1. On the S-PLUS menu bar, make the following menu choices: **EnvironmentalStats>Probability Distributions and Random Numbers>Density, CDF, Quantiles**. This will bring up the Densities, Cumulative Probabilities, or Quantiles dialog box.

2. For the Data to Use buttons, choose **Expression**. In the Expression box, type **0, 0.5, 1, 1.5, 2**. In the Distribution box, choose **Lognormal (Alternative)**. In the mean box type **0.6**. In the cv box type **0.5**. Under the Probability or Quantile group, **check** the Cumulative Probability box. Click **OK** or **Apply**.

Command

To produce the lognormal cdf shown in Figure 4.4, type this command.

```
> cdfplot(distribution="lnorm.alt",
    param.list=list(mean=0.6, cv=0.5))
```

To compute the value of this lognormal cdf at 0, 0.5, 1, 1.5, and 2, type this command.

```
> plnorm.alt(c(0, 0.5, 1, 1.5, 2), mean=0.6, cv=0.5)
```

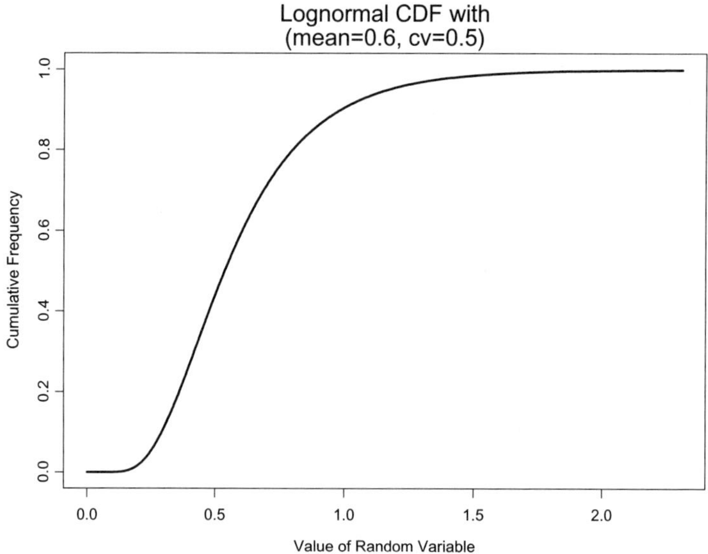

Figure 4.4. A lognormal cumulative distribution function.

4.4   Quantiles and Percentiles

Loosely speaking, the pth *quantile* of a population is the (a) number such that a fraction p of the population is less than or equal to this number. The pth quantile is the same as the $100p^{th}$ *percentile*; for example, the 0.5 quantile is the same as the 50th percentile.

Here is a more technical definition of a quantile. If X is a random variable with some specified distribution, the pth *quantile* of the distribution of X, denoted x_p, is a (the) number that satisfies:

$$\Pr\left(X < x_p\right) \leq p \leq \Pr\left(X \leq x_p\right) \tag{4.5}$$

where p is a number between 0 and 1 (inclusive). If there is more than one number that satisfies the above condition, the pth quantile of X is often taken to be the average of the smallest and largest numbers that satisfy the condition. The S-PLUS functions for computing quantiles, however, return the smallest number that satisfies the above condition.

If X is a continuous random variable, the pth quantile of X is simply defined as the value such that the cdf of that value is equal to p:

$$\Pr\left(X \leq x_p\right) = F\left(x_p\right) = p \tag{4.6}$$

The $100p^{th}$ *percentile* is another name for the pth quantile. That is, the $100p^{th}$ percentile is the (a) number such that $100p\%$ of the distribution lies below this number.

4.4.1   Quantiles for Lognormal Distribution

A plot of the cumulative distribution function makes it easy to visually pick out important quantiles, such as the median (50th percentile) or the 95th percentile. Looking at the cdf of the lognormal distribution shown in Figure 4.4, the median (50th percentile) is about 0.5 and the 95th percentile is about 1.1 (the actual values are 0.537 and 1.167, respectively).

Menu

To compute the 50th and 95th percentiles of the lognormal distribution shown in Figure 4.4, follow these steps.

1. On the S-PLUS menu bar, make the following menu choices: **EnvironmentalStats>Probability Distributions and Random Numbers>Density, CDF, Quantiles**. This will bring up the Densities, Cumulative Probabilities, or Quantiles dialog box.

2. For the Data to Use buttons, choose **Expression**. In the Expression box, type **0.5, 0.95**. In the Distribution box, choose **Lognormal (Alternative)**. In the mean box type **0.6**. In the cv box type **0.5**. Under the Probability or Quantile group, **check** the Quantile box. Click **OK** or **Apply**.

Command

To compute the 50th and 95th percentiles of the lognormal distribution shown in Figure 4.4 type this command.

```
> qlnorm.alt(c(0.5, 0.95), mean=0.6, cv=0.5)
```

4.5   Generating Random Numbers

With the advance of modern computers, experiments and simulations that just a decade ago would have required an enormous amount of time to complete using large-scale computers can now be easily carried out on personal computers. Simulation is fast becoming an important tool in environmental statistics and all fields of statistics in general.

For all of the distributions shown in Figure 4.1, you can generate random numbers (actually, pseudo-random numbers) from these distributions using S-PLUS and ENVIRONMENTALSTATS for S-PLUS. See Chapter 9 for a discussion of how random numbers are generated in S-PLUS and ENVIRONMENTALSTATS for S-PLUS.

4.5.1   Generating Random Numbers from a Univariate Distribution

Five random numbers from the lognormal distribution shown in Figure 4.3 and Figure 4.4 are: 0.1903606, 0.2733133, 0.6742302, 0.7006088, and 0.6292198.

Menu

To generate these five random numbers, follow these steps.

1. On the S-PLUS menu bar, make the following menu choices: **EnvironmentalStats>Probability Distributions and Random Numbers>Random Numbers>Univariate**. This will bring up the Univariate Random Number Generation dialog box.
2. In the Sample Size box, type **5**. For Set Seed with type **23**. In the Distribution box, choose **Lognormal (Alternative)**. In the mean box type **0.6**. In the cv box type **0.5**. Click **OK** or **Apply**.

Note that if you leave the Set Seed with box blank, the random numbers you generate will not be the same.

Command

To generate these five random numbers, type these commands.

```
> set.seed(23)
> rlnorm.alt(5, mean=0.6, cv=0.5)
```

Note that if you leave out the call to `set.seed`, the random numbers you generate will not be the same.

4.5.2   Generating Multivariate Normal Random Numbers

In S-PLUS you can generate random observations from a multivariate normal distribution using the function `rmvnorm`. Although you cannot perform a similar action from the S-PLUS menu, you can generate such observations from the ENVIRONMENTALSTATS for S-PLUS menu. Consider a bivariate normal distribution with the following parameters:

$$\mu = (5, 10)$$

$$\sigma = (1, 2) \tag{4.7}$$

$$\rho = \begin{pmatrix} 1 & 0.5 \\ 0.5 & 1 \end{pmatrix}$$

where μ denotes the means, and σ denotes the standard deviations, and ρ denotes the correlation matrix. Here are three random observations from this bivariate distribution:

```
          [,1]         [,2]
[1,]  4.532853   9.103044
[2,]  6.007556   8.343978
[3,]  5.490049  10.609173
```

Menu

To generate these three bivariate observations, follow these steps.

1. On the S-PLUS menu bar, make the following menu choices: **EnvironmentalStats>Probability Distributions and Random Numbers>Random Numbers>Multivariate>Normal**. This will bring up the MV Normal Random Number Generation dialog box.
2. For Sample Size type **3**. For Set Seed with type **47**. For Number of Vars select **2**. For Means type **5, 10**. For Supply select **Correlations/SDs**. For Correlation Matrix type **matrix(c(1, 0.5, 0.5, 1), ncol=2)**. For Standard Deviations type **1, 2**. Click **OK** or **Apply**.

Command

To generate these three bivariate observations, type these commands.

```
> set.seed(47)
> rmvnorm(3, mean=c(5, 10), sd=c(1, 2),
    cov=matrix(c(1, 0.5, 0.5, 1), ncol=2))
```

Note that the matrix we supply as the argument `cov` is actually the correlation matrix (see the S-PLUS help file *MVNormal*).

4.5.3   Generating Multivariate Observations Based on Rank Correlations

In ENVIRONMENTALSTATS for S-PLUS, you can generate multivariate correlated observations where each variable has an arbitrary distribution. For example, you can generate a multivariate observation ($X1$, $X2$) where $X1$ comes from a normal distribution and $X2$ comes from a lognormal distribution. Suppose $X1$ follows a normal distribution with mean 5 and standard deviation 1, $X2$ follows a lognormal distribution with mean 10 and CV 2, and we desire a rank correlation specified by ρ in Equation (4.7). Here are three random observations from this bivariate distribution:

```
            X1          X2
[1,]  5.299270  10.310089
[2,]  3.799289   4.921048
[3,]  6.368534   7.034561
```

Menu

To generate these three bivariate observations, follow these steps.

1. On the S-PLUS menu bar, make the following menu choices: **EnvironmentalStats>Probability Distributions and Random Numbers>Random Numbers>Multivariate>Based on Rank Correlation**. This will bring up the Multivariate Random Number Generation dialog box.
2. For Sample Size type **3**. For Set Seed with type **47**. For Number of Vars select **2**. For Correlation Matrix type **matrix(c(1, 0.5, 0.5, 1), ncol=2)**.
3. Click on the **Variable 1** tab. For Variable Name type **X1**. For Distribution Type select **Parametric**. For Sampling Method choose **Random**. For Distribution select **Normal**, for mean type **5**, and for sd type **1**.
4. Click on the **Variable 2** tab. For Variable Name type **X2**. For Distribution Type select **Parametric**. For Sampling Method choose **Random**. For Distribution select **Lognormal (alternative)**, for mean type **10**, and for cv type **2**. Click **OK** or **Apply**.

Command

To generate these three bivariate observations, type this command.

```
> simulate.mv.matrix(n=3,
    distributions=c(X1="norm", X2="lnorm.alt"),
    param.list=list(X1=list(mean=5, sd=1),
                    X2=list(mean=10, cv=2)),
    cor.mat=matrix(c(1, 0.5, 0.5, 1), ncol=2), seed=105)
```

4.6 Summary

- Figure 4.1 displays examples of the probability density functions for the probability distributions available in S-PLUS and ENVIRONMENTALSTATS for S-PLUS. Some of these distributions are already available in S-PLUS, and some have been added or modified in ENVIRONMENTALSTATS for S-PLUS.
- Table 4.1 lists the probability distributions available in S-PLUS and ENVIRONMENTALSTATS for S-PLUS, along with their abbreviations and associated parameters. For each of these distributions, you can compute the probability density function (pdf), the cumulative distribution function (cdf), quantiles, and random numbers. You can also plot the pdf and/or cdf.
- Table 4.2 lists the menu items and functions available in ENVIRONMENTALSTATS for S-PLUS for plotting probability distributions, computing quantities associated with these distributions, and generating random numbers from these distributions.

5

Estimating Distribution Parameters and Quantiles

5.1 Introduction

In Chapter 2 we discussed the ideas of a population and a sample. Chapter 4 described probability distributions, which are used to model populations. Based on using the graphical tools discussed in Chapter 3 to look at your data, and based on your knowledge of the mechanism producing the data, you can model the data from your sampling program as having come from a particular kind of probability distribution. Once you decide on what probability distribution to use (if any), you usually need to estimate the parameters associated with that distribution. For example, you may need to compare the mean or 95th percentile of the concentration of a chemical in soil, groundwater, surface water, or air with some fixed standard. This chapter discusses the menu items and functions available in ENVIRONMENTALSTATS for S-PLUS for estimating distribution parameters and quantiles for various probability distributions (as well as constructing confidence intervals for these quantities). See Chapter 5 of Millard and Neerchal (2001) for a more in-depth discussion of this topic.

Table 5.1 lists the menu items and functions available in ENVIRONMENTALSTATS for S-PLUS for estimating distribution parameters and quantiles. All menu items fall under the menu choice **EnvironmentalStats**. For example, the first entry in the table describes what you can do if you make the selection **EnvironmentalStats>Estimation>Parameters**.

Menu Item	Functions	Description
Estimation>Parameters	`eabb`	Estimate the parameters of the distribution with abbreviation *abb*, and optionally construct a confidence interval for the parameters.

Estimation>Quantiles	`eqnorm`
	`eqlnorm`
	`eqpois`
	`eqnpar`
		Estimate a quantile for a normal, lognormal, or Poisson distribution, and optionally construct a confidence interval for the quantile.

Table 5.1. Menu items and functions in ENVIRONMENTALSTATS for S-PLUS for estimating distribution parameters and quantiles.

5.2   Estimating Distribution Parameters

Table 4.1 lists the probability distributions available in S-PLUS and ENVIRONMENTALSTATS for S-PLUS. For most of these distributions, there is a menu item and functions for estimating the parameters of these distributions. (See the help file *Estimating Distribution Parameters and Quantiles* for a complete list.) The form of the names of the functions is e*abb*, where *abb* denotes the abbreviation of the distribution name (see column 2 of Table 4.1). For example, the function enorm estimates the mean and standard deviation based on a set of observations assumed to come from a normal distribution, and also optionally allows you to construct a confidence interval for the mean or variance.

5.2.1   Estimating Parameters of a Normal Distribution

Recall that in Chapter 1 we saw that the Reference area TcCB data appeared to come from a lognormal distribution. Using ENVIRONMENTALSTATS for S-PLUS, we can estimate the mean and standard deviation of the log-transformed data. These estimates are approximately –0.62 and 0.47 log(ppb). Figure 5.1 shows a density histogram of the log-transformed Reference area TcCB data, along with the fitted normal distribution based on these estimates. The two-sided 95% confidence interval for the mean is [-0.76, -0.48] log(ppb).

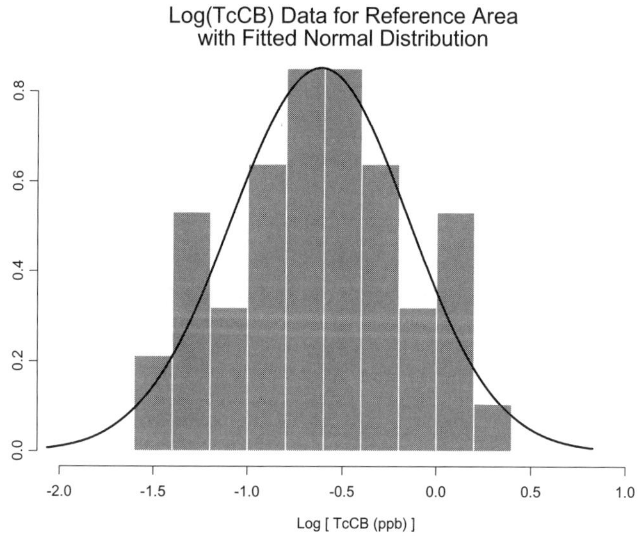

Figure 5.1. Histogram of log-transformed Reference area TcCB data with fitted normal distribution.

Menu

To estimate the mean and standard deviation of the log-transformed Reference area TcCB concentrations and compute a 95% confidence interval for the mean using the ENVIRONMENTALSTATS for S-PLUS pull-down menu, follow these steps.

1. In the Object Explorer, find and highlight **new.epa.94b.tccb.df**.
2. On the S-PLUS menu bar, make the following menu choices: **EnvironmentalStats>Estimation>Parameters**. This will bring up the Estimate Distribution Parameters dialog box.
3. For Data to Use, select **Pre-Defined Data**. In the Data Set box, select **new.epa.94b.tccb.df**. In the Variable box select **log.TcCB**. In the Subset Rows with box type **Area=="Reference"**. In the Distribution/Estimation section, make sure **Normal** is selected in the Distribution box, and **mvue** is selected in the Estimation Method box. Under the Confidence Interval section, **check** the Confidence Interval box. In the CI Type box, select **two-sided**. In the CI Method box, select **exact**. In the Confidence Level (%) box, select **95**. In the Parameter box, select **mean**. Click **OK** or **Apply**.

Command

To estimate the mean and standard deviation of the log-transformed Reference area TcCB concentrations and create a 95% confidence interval for the mean, type these commands.

```
> attach(epa.94b.tccb.df)
> enorm(log(TcCB[Area=="Reference"]), ci=T,
    conf.level=0.95)
> detach()
```

5.2.2   Estimating Parameters of a Lognormal Distribution

Rather than estimate parameters based on the log-transformed TcCB Reference area data, we can estimate the parameters of the lognormal distribution based on the original scale. For the untransformed Reference area TcCB data, the estimated mean is 0.6 ppb and the estimated coefficient of variation is 0.49. Figure 5.2 shows a density histogram of the Reference area TcCB data, along with the fitted lognormal distribution based on these estimates. The two-sided 95% confidence interval for the mean based on Land's method is [0.52, 0.7] ppb.

Menu

To estimate the mean and coefficient of variation of the Reference area TcCB concentrations and compute a 95% confidence interval for the mean using the ENVIRONMENTALSTATS for S-PLUS pull-down menu, follow these steps.

1. In the Object Explorer, find and highlight **epa.94b.tccb.df**.
2. On the S-PLUS menu bar, make the following menu choices: **EnvironmentalStats>Estimation>Parameters**. This will bring up the Estimate Distribution Parameters dialog box.
3. For Data to Use, select **Pre-Defined Data**. In the Data Set box, select **epa.94b.tccb.df**. In the Variable box select **TcCB**. In the Subset Rows with box type **Area=="Reference"**. For Distribution select **Lognormal (alternative)**. For Estimation Method select **mvue**. Under the Confidence Interval section, **check** the Confidence Interval box. In the CI Type box, select **two-sided**. In the CI Method box, select **land**. In the Confidence Level (%) box, select **95**. Click **OK** or **Apply**.

Command

To estimate the mean and standard deviation of the Reference area TcCB concentrations and create a 95% confidence interval for the mean, type these commands.

```
> attach(epa.94b.tccb.df)
> elnorm.alt(TcCB[Area=="Reference"], ci=T,
    conf.level=0.95)
> detach()
```

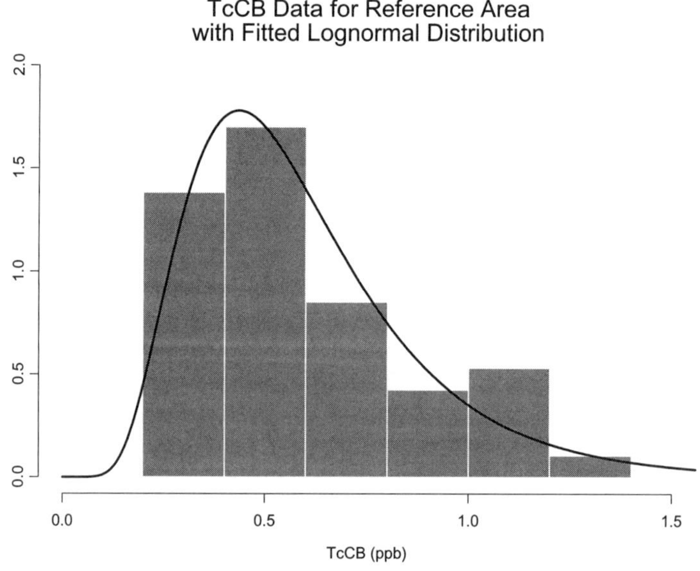

Figure 5.2. Histogram of Reference area TcCB data with fitted lognormal distribution.

5.2.3   Estimating the Parameter of a Binomial Distribution

The guidance document *Statistical Analysis of Ground-Water Monitoring Data at RCRA Facilities: Addendum to Interim Final Guidance* (USEPA, 1992c, p. 36) contains observations on benzene concentrations (ppb) in groundwater from six background wells sampled monthly for 6 months. The data are stored in the data frame epa.92c.benzene1.df in ENVIRONMENTALSTATS for S-PLUS.

```
> epa.92c.benzene1.df
   Benzene.orig Benzene Censored Month Well
1            <2        2        T     1    1
2            <2        2        T     2    1
3            <2        2        T     3    1
4            <2        2        T     4    1
5            <2        2        T     5    1
6            <2        2        T     6    1
 .
31           <2        2        T     1    6
32           <2        2        T     2    6
33           <2        2        T     3    6
34           <2        2        T     4    6
35           10       10        F     5    6
36           <2        2        T     6    6
```

Nondetect values are reported as "<2." Of the 36 values, 33 are nondetects. Based on these data, the probability of observing a nondetect value at any of the six wells is estimated to be about 92%. The exact two-sided 95% confidence interval for this binomial proportion is [0.78, 0.98].

Menu

To estimate the probability of observing a nondetect value and construct a confidence interval for this probability, follow these steps.

1. In the Object Explorer, find and highlight the data frame **epa.92c.benzene1.df**.
2. On the S-PLUS menu bar, make the following menu choices: **EnvironmentalStats>Estimation>Parameters**. This will bring up the Estimate Distribution Parameters dialog box.
3. For Data to Use, select **Pre-Defined Data**. In the Data Set box, select **epa.92c.benzene1.df**. In the Variable box, select **Censored**. For Distribution select **Binomial**. **Check** the Confidence Interval box. In the CI Type box, select **two-sided**. In the CI Method box, select **exact**. In the Confidence Level (%) box, select **95**. Click **OK** or **Apply**.

Command

To estimate the probability of observing a nondetect value and construct a confidence interval for this probability, type these commands.

```
> attach(epa.92c.benzene1.df)
> ebinom(Censored, ci=T, conf.level=0.95)
> detach()
```

5.3   Estimating Distribution Quantiles

We defined the pth quantile, or the $100p^{th}$ percentile, of a distribution in Section 4.4 of Chapter 4. Quantiles or percentiles are sometimes used in environmental standards and regulations (e.g., Berthouex and Brown, 1994, p. 65). For example, in order to determine compliance, you may be required to estimate an extreme percentile (e.g., the 95th percentile) for the "background level" distribution, and then compare observations at compliance wells or remediated areas to this upper percentile (or an upper confidence limit for this percentile). In the context of soil cleanup, USEPA (1994b, pp. 4.8–4.9) has called this the "Hot-Measurement Comparison." (There are some major problems with this technique that are discussed in Chapters 6 and 7 of Millard and Neerchal (2001).)

As another example, when monitoring groundwater around a RCRA landfill, the site may be in "compliance monitoring" with a fixed compliance limit for a particular chemical. The fixed compliance limit may be a maximum concentration limit (MCL) or an alternate concentration limit (ACL). Most MCLs in fact represent average levels, but sometimes an MCL or ACL may represent a limit that should be exceeded only a small fraction of the time, for example, the 99th percentile of the distribution. In this case you need to compare the 99th percentile of the distribution of the chemical's concentration in the groundwater with the MCL. If the estimated 99th percentile (or a lower confidence limit for this percentile) is greater than the MCL, then you are out of compliance.

For the normal, lognormal, and Poisson distributions, there is a menu item and functions for estimating quantiles of the distribution and optionally constructing confidence intervals for these quantiles (see Table 5.1). The functions for doing this are `eqnorm`, `eqlnorm`, and `eqpois`. You can also estimate quantiles and create confidence intervals for them nonparametrically using the function `eqnpar`.

5.3.1   Estimating Quantiles of a Normal Distribution

The guidance document *Statistical Analysis of Ground-Water Monitoring Data at RCRA Facilities, Interim Final Guidance* (USEPA, 1989b) contains an example on pages 6-11 to 6-13 of aldicarb concentrations (ppm) at three groundwater monitoring compliance wells (four monthly samples at each well). In ENVIRONMENTALSTATS for S-PLUS, these data are stored in the data frame `epa.89b.aldicarb2.df`.

```
> epa.89b.aldicarb2.df
   Aldicarb Month Well
 1     19.9     1    1
 2     29.6     2    1
 3     18.7     3    1
 4     24.2     4    1
 5     23.7     1    2
 6     21.9     2    2
 7     26.9     3    2
 8     26.1     4    2
 9     25.6     1    3
10     23.3     2    3
11     22.3     3    3
12     26.9     4    3
```

In this example, it is assumed that the permit establishes an ACL of 50 ppm that should not be exceeded more than 5% of the time. Thus, for each well, we need to determine whether the 95th percentile of aldicarb concentrations is greater than the ACL of 50 ppm. To do this, we need to compute a one-sided *lower* confidence limit for the 95th percentile. If this confidence limit is above the ACL, then we can be fairly sure that the true 95th percentile is above the ACL. (Note: the EPA guidance document incorrectly computes a one-sided *upper* confidence limit for the 95th percentile.)

Table 5.2 displays the estimated mean, standard deviation, 95th percentile, and lower one-sided 99% confidence limit for the 95th percentile for each of the three compliance wells. All of the lower confidence limits for the 95th percentile are well below the ACL of 50 ppm, so none of the wells is out of compliance. To give you an idea of the amount of variability in these estimates of the 95th percentile, we can also create two-sided confidence intervals. The two-sided 99% confidence intervals for the 95th percentiles at each well (rounded to whole numbers) are: [25, 80], [25, 51], and [25, 49]. Of course, a major problem here is the very small sample size ($n = 4$) at each well.

Statistic	Well 1	Well 2	Well 3
Mean	23.1	24.7	24.5
SD	4.9	2.3	2.1
95th Percentile	31.2	28.4	28.0
99% LCL on 95th Percentile	25.3	25.7	25.5

Table 5.2. Statistics for aldicarb concentrations at three compliance wells.

Menu

To produce the one-sided lower 99% confidence limits for the 95th percentiles for each of the compliance wells using the ENVIRONMENTALSTATS for S-PLUS pull-down menu, follow these steps.

1. In the Object Explorer, find the data frame **epa.89b.aldicarb2.df**.
2. On the S-PLUS menu bar, make the following menu choices: **EnvironmentalStats>Estimation>Quantiles**. This will bring up the Estimate Distribution Quantiles dialog box.
3. For Data to Use, select **Pre-Defined Data**. For Data Set box, select **epa.89b.aldicarb2.df**. For Variable select **Aldicarb**. In the Subset Rows with box, type **Well=="1"**. In the Quantile(s) box, select **0.95**.
4. Under the Distribution/Estimation group, for Type, make sure the **Parametric** button is selected. In the Distribution box, select **Normal**. For Estimation Method, select **qmle**.
5. Under the Confidence Interval group, make sure the Confidence Interval box is **checked**. For CI Type, select **lower**. For CI Method, select **exact**. For Confidence Level (%), select **99**. Click **OK** or **Apply**.

This will produce the estimated 95th percentile and one-sided lower 99% confidence limit for this percentile for the first well. To produce these quantities for the other two wells, in Step 3 above, in the Subset Rows with box, type **Well=="2"** or **Well=="3"**.

Command

To produce the one-sided lower 99% confidence limits for the 95th percentiles for each of the compliance wells, type these commands.

```
> attach(epa.89b.aldicarb2.df)
> eqnorm(Aldicarb[Well=="1"], p=0.95, ci=T,
    ci.type="lower", conf.level=0.99)
> eqnorm(Aldicarb[Well=="2"], p=0.95, ci=T,
    ci.type="lower", conf.level=0.99)
> eqnorm(Aldicarb[Well=="3"], p=0.95, ci=T,
    ci.type="lower", conf.level=0.99)
> detach()
```

Instead of calling `eqnorm` three separate times, you could instead just use the following single command:

```
> lapply(split(Aldicarb, Well), eqnorm, p=0.95, ci=T,
    ci.type="lower", conf.level=0.99)
```

5.3.2   Estimating Quantiles of a Lognormal Distribution

The guidance document USEPA (1992c, pp. 52–54) contains an example of chrysene concentrations (ppb) at five groundwater monitoring compliance wells (four monthly samples at each well). In ENVIRONMENTALSTATS for S-PLUS these data are stored in the data frame `epa.92c.chrysene.df`. They are assumed to come from a lognormal distribution.

```
> epa.92c.chrysene.df
   Chrysene Month Well
1      19.7      1    1
2      39.2      2    1
3       7.8      3    1
4      12.8      4    1
 .
17     47.0      1    5
18     30.5      2    5
19     15.0      3    5
20     23.4      4    5
```

In this example, it is assumed that the permit establishes an ACL of 80 ppb that should not be exceeded more than 5% of the time. Thus, for each well, we need to determine whether the 95th percentile of chrysene concentrations is greater than the ACL of 80 ppb. To do this, we need to compute a one-sided *lower* confidence limit for the 95th percentile. If this confidence limit is above the ACL, then we can be fairly sure that the true 95th percentile is above the ACL. (Note: The EPA guidance document incorrectly computes a one-sided *upper* confidence limit for the 95th percentile.)

Table 5.3 displays the estimated mean, coefficient of variation, 95th percentile, and lower one-sided 99% confidence limit for the 95th percentile for each of the five compliance wells. All of the lower confidence limits for the 95th percentile are well below the ACL of 80 ppb, so none of the wells is out of compliance. To give you an idea of the amount of variability in these estimates of the 95th percentile, we can also create two-sided confidence intervals. The two-sided 99% confidence intervals for the 95th percentiles at each well (rounded to whole numbers) are: [25, 178], [11, 63], [52, 235], [27, 95], and [34, 185]. As in the previous example, a major problem here is the very small sample size ($n = 4$) at each well.

Statistic	Well 1	Well 2	Well 3	Well 4	Well 5
Mean	19.9	9.8	46.3	24.7	29.0
CV	0.7	0.5	0.4	0.2	0.5
95th Percentile	51.4	19.0	78.5	36.5	58.5
99% LCL on 95th Percentile	22.6	11.1	51.6	26.9	32.9

Table 5.3. Statistics for chrysene concentrations at five compliance wells.

Menu

To produce the one-sided lower 99% confidence limits for the 95th percentiles for each of the compliance wells follow these steps.

1.   In the Object Explorer find the data frame **epa.92c.chrysene.df**.

2. On the S-PLUS menu bar, make the following menu choices: **EnvironmentalStats>Estimation>Quantiles**. This will bring up the Estimate Distribution Quantiles dialog box.

3. For Data to Use, select **Pre-Defined Data**. For Data Set box, select **epa.92c.chrysene.df**. For Variable select **Chrysene**. In the Subset Rows with box, type **Well=="1"**. In the Quantile(s) box, select **0.95**.

4. Under the Distribution/Estimation group, for Type, make sure the **Parametric** button is selected. In the Distribution box, select **Lognormal**. For Estimation Method, select **qmle**.

5. Under the Confidence Interval group, make sure the Confidence Interval box is **checked**. For CI Type, select **lower**. For CI Method, select **exact**. For Confidence Level (%), select **99**. Click **OK** or **Apply**.

This will produce the estimated 95th percentile and one-sided lower 99% confidence limit for this percentile for the first well. To produce these quantities for the other four wells, in Step 3 above, in the Subset Rows with box, type **Well=="2"**, **Well=="3"**, **Well=="4"**, or **Well=="5"**.

Command

To produce the one-sided lower 99% confidence limits for the 95th percentiles for each of the compliance wells type these commands.

```
> attach(epa.92c.chrysene.df)
> eqlnorm(Chrysene[Well=="1"], p=0.95, ci=T,
    ci.type="lower", conf.level=0.99)
> eqlnorm(Chrysene[Well=="2"], p=0.95, ci=T,
    ci.type="lower", conf.level=0.99)
> eqlnorm(Chrysene[Well=="3"], p=0.95, ci=T,
    ci.type="lower", conf.level=0.99)
> eqlnorm(Chrysene[Well=="4"], p=0.95, ci=T,
    ci.type="lower", conf.level=0.99)
> eqlnorm(Chrysene[Well=="5"], p=0.95, ci=T,
    ci.type="lower", conf.level=0.99)
> detach()
```

Instead of calling `eqlnorm` five separate times, you could instead just use the following single command:

```
> lapply(split(Chrysene, Well), eqlnorm, p=0.95, ci=T,
    ci.type="lower", conf.level=0.99)
```

5.3.3   Nonparametric Estimates of Quantiles

To estimate quantiles nonparametrically, all you need to do is estimate the cdf nonparametrically using the empirical cdf, then use linear interpolation (if necessary). Graphically, this just means connecting the points in the quantile plot by straight lines, finding the value of p on the y-axis, and determining the corresponding number on the x-axis. For example, looking at Figure 1.3 on page 13, we might estimate the median (i.e., $p = 0.5$) of the Reference area TcCB data to be about 0.5 (the actual value is 0.54).

One problem with estimating quantiles nonparametrically versus parametrically is that you need many more observations to estimate extreme quantiles with good precision. In fact, even though S-PLUS will give us estimates for the 5th and 95th percentiles, it does not make sense intuitively that we should be able to estimate anything less than the 10th percentile or anything more than the 90th percentile with any kind of precision if $n = 10$. This characteristic becomes clear when we create nonparametric confidence intervals for quantiles. On the other hand, an advantage to estimating quantiles nonparametrically is that it is often easy to deal with censored values since all you have to do is rank them.

Nonparametric confidence intervals for quantiles are based on the ranked data. For example, Table 5.4 illustrates the confidence levels associated with a one-sided upper confidence interval for the 95th percentile, based on various sample sizes, assuming the upper confidence limit is the maximum value. You can see that a confidence level greater than 95% cannot be achieved until the sample size is larger than $n = 50$. See Chapter 5 of Millard and Neerchal (2001) for a detailed discussion of estimating quantiles nonparametrically and constructing nonparametric confidence intervals for quantiles.

Sample Size (n)	Confidence Level (%)
5	23
10	40
15	54
20	64
25	72
50	92
75	98
100	99

Table 5.4. Confidence levels for one-sided upper nonparametric confidence interval for the 95th percentile, based on using the maximum value as the upper confidence limit.

The guidance document *Statistical Analysis of Ground-Water Monitoring Data at RCRA Facilities: Addendum to Interim Final Guidance* (USEPA, 1992c, pp. 55–56) contains an example of copper concentrations (ppb) at five groundwater monitoring wells: three background wells and two compliance wells (eight monthly samples at each well, except the first four missing at the two compliance wells). These data are shown in Table 5.5 and in ENVIRONMENTALSTATS for S-PLUS they are stored in the data frame `epa.92c.copper2.df`. Note that 15 out of the 24 observations at the background wells are nondetects recorded as "<5."

	Background			Compliance	
Month	**Well 1**	**Well 2**	**Well 3**	**Well 4**	**Well 5**
1	<5	9.2	<5		
2	<5	<5	5.4		
3	7.5	<5	6.7		
4	<5	6.1	<5		
5	<5	8	<5	6.2	<5
6	<5	5.9	<5	<5	<5
7	6.4	<5	<5	7.8	5.6
8	6	<5"	<5	10.4	<5

Table 5.5. Copper concentrations (ppb) at five groundwater monitoring wells.

USEPA (1992c, pp. 55–56) describes how to compute a nonparametric upper confidence limit for an extreme percentile of the copper concentrations at the three background wells, with the idea that this limit will be used as a threshold value for concentrations observed in the compliance wells (i.e., if the concentration at a compliance well exceeds this limit, this indicates there may be contamination in the groundwater). This is the "Hot-Measurement Comparison," and, as already stated, there are some major problems with this technique that are discussed in Chapters 6 and 7 of Millard and Neerchal (2001).

For this data set, $15/24 = 62.5\%$ of the values at the background wells are nondetects, so the estimated median is some value less than 5. The estimated 95th percentile is 7.925, and using the largest value of 9.2 ppb as the upper confidence limit for the 95th percentile yields a one-sided upper confidence interval with a confidence level of only 71%! If we are willing to use 9.2 as a threshold value, then Well 4 indicates possible contamination, whereas Well 5 does not.

Menu

To nonparametrically estimate the 95th percentile of copper concentrations at the background wells, and produce the one-sided upper confidence limit for the 95th percentile, follow these steps.

1.   In the Object Explorer, find the data frame **epa.92c.copper2.df**.
2.   On the S-PLUS menu bar, make the following menu choices: **EnvironmentalStats>Estimation>Quantiles**. This will bring up the Estimate Distribution Quantiles dialog box.
3.   For Data to Use, select **Pre-Defined Data**. For Data Set box, select **epa.92c.copper2.df**. For Variable select **Copper**. In the Subset Rows with box, type **Well.type=="Background"**. In the Quantile(s) box, select **0.95**.
4.   Under the Distribution/Estimation group, for Type, select **Nonparametric**.

5. Under the Confidence Interval group, make sure the Confidence Interval box is **checked**. For CI Type, select **upper**. For CI Method, select **exact**. For Confidence Level (%), type **95**. For Lower Bound, type **0**. Click **OK** or **Apply**.

Command

To nonparametrically estimate the 95th percentile of the copper concentrations at the background wells, and produce the one-sided upper confidence limit for the 95th percentile, type these commands.

```
> attach(epa.92c.copper2.df)
> eqnpar(Copper[Well.type=="Background"], p=0.95, ci=T,
    lb=0, ci.type="upper", approx.conf.level=0.95)
> detach()
```

5.4 Summary

- Whether you are conducting a preliminary, descriptive study of the environment or monitoring the environment for contamination under a specific regulation, you usually need to characterize the distribution of whatever you are looking at (e.g., a chemical in the environment), which involves *estimating distribution parameters* such as the mean, median, standard deviation, 95th percentile, etc.
- Table 5.1 shows the menu items and functions available in ENVIRONMENTALSTATS for S-PLUS for estimating distribution parameters and quantiles, and optionally constructing confidence intervals for these quantities.
- One problem with estimating quantiles nonparametrically versus parametrically is that you need many more observations to estimate extreme quantiles with good precision.

6

Prediction and Tolerance Intervals

6.1 Introduction

Any activity that requires constant monitoring over time and the comparison of new values to "background" or "standard" values creates a decision problem: if the new values greatly exceed the background values, has a change really occurred, or have the true underlying concentrations stayed the same and this is just a "chance" event? Statistical tests are used as objective tools to decide whether a change has occurred (although the choice of Type I error level and acceptable power are subjective decisions). For a monitoring program that involves numerous tests over time, figuring out how to balance the **overall** Type I error with the power of detecting a change is not a trivial problem, but it is also a problem that has been dealt with for a long time in the statistical literature under the heading of "multiple comparisons" (see the help file *Glossary: Multiple Comparisons*). Prediction intervals and tolerance intervals are two tools that you can use to attempt to solve the multiple comparisons problem. This chapter discusses the menu items and functions available in ENVIRONMENTALSTATS for S-PLUS for constructing prediction and tolerance intervals. See Chapter 6 of Millard and Neerchal (2001) for a more in-depth discussion of this topic.

6.2 Prediction Intervals

A **prediction interval** for some population is an interval on the real line constructed so that it will contain k future observations or averages from that population with some specified probability $(1-\alpha)100\%$, where α is some fraction between 0 and 1 (usually α is less than 0.5), and k is some pre-specified positive integer. Just as for confidence intervals, the quantity $(1-\alpha)100\%$ is called the **confidence coefficient** or **confidence level** associated with the prediction interval.

Table 6.1 lists the functions available in ENVIRONMENTALSTATS for S-PLUS for constructing prediction intervals. From the S-PLUS menu bar, you can construct prediction intervals by making the selection **Environmental-Stats>Estimation>Prediction Intervals**.

Distribution	Functions for Prediction Intervals	Description
Normal	`pred.int.norm`	Construct a prediction interval for the next k observations or next k means from a normal distribution.
	`pred.int.norm.K`	Compute the value of K for a prediction interval for a normal distribution
	`pred.int.norm.simultaneous`	Construct a simultaneous prediction interval for the next r sampling occasions based on a normal distribution.
	`pred.int.norm.simultaneous.K`	Compute the value of K for a simultaneous prediction interval for the next r sampling occasions based on a normal distribution.
Lognormal	`pred.int.lnorm`	Construct a prediction interval for the next k observations or next k geometric means from a lognormal distribution.
	`pred.int.lnorm.simultaneous`	Construct a simultaneous prediction interval for the next r sampling occasions based on a lognormal distribution.
Poisson	`pred.int.pois`	Construct a prediction interval for the next k observations or sums from a Poisson distribution.
Nonparametric	`pred.int.npar`	Construct a nonparametric prediction interval for the next k of m observations.

Table 6.1. Functions in ENVIRONMENTALSTATS for S-PLUS for constructing prediction intervals.

The basic idea of a prediction interval is to assume a particular probability distribution (e.g., normal, lognormal, etc.) for some process generating the data (e.g., quarterly observations of chemical concentrations in groundwater), compute sample statistics from a baseline sample, and then use these sample statistics to construct a prediction interval, assuming the distribution of the data does not change in the future. For example, if X denotes a random variable from some population, and we know what the population looks like (e.g., N(10, 2)) so we can compute the quantiles of the population, then a $(1-\alpha)100\%$ two-sided prediction interval for the next $k = 1$ observation of X is given by:

$$\left[x_{\alpha/2} , \ x_{1-\alpha/2} \right] \tag{6.1}$$

where x_p denotes the pth quantile of the distribution of X. Similarly, a $(1-\alpha)100\%$ one-sided upper prediction interval for the next observation is given by:

$$\left[-\infty \, , \ x_{1-\alpha} \right] \tag{6.2}$$

and a $(1-\alpha)100\%$ one-sided lower prediction interval for the next observation is given by:

$$\left[x_\alpha \, , \ \infty \right] \tag{6.3}$$

See Chapter 6 of Millard and Neerchal (2001) for the corresponding equations for general values of k.

Usually the true distribution of X is unknown, so the values of the prediction limits have to be estimated based on estimating the parameters of the distribution of X. For the usual case when the exact distribution of X is unknown, a prediction interval is thus a random interval; that is, the lower and upper bounds are random variables computed based on sample statistics in the baseline sample. Prior to taking one specific baseline sample, the probability that the prediction interval *will* contain the next k observations is $(1-\alpha)100\%$. Once a specific baseline sample is taken and the prediction interval based on that sample is computed, the probability that that prediction interval will contain the next k observations is not necessarily $(1-\alpha)100\%$, but it should be close to this value for a moderately large sample size.

Suppose an experiment is performed N times, and suppose that for each experiment:

1. A sample is taken and a $(1-\alpha)100\%$ prediction interval for $k = 1$ future observation is computed.
2. One future observation is generated and compared to the prediction interval.

Then the number of times a prediction interval generated in Step 1 above will contain a future observation generated in step 2 above is a binomial random variable with parameters $n = N$ and $p = 1-\alpha$, that is, it follows a B(N, $1-\alpha$) distribution. Figure 6.1 shows the results of such a simulated experiment in which a random sample of $n = 10$ observations was taken from a N(5, 1) distribution and an 80% prediction interval for $k = 1$ future observation was constructed based on these 10 observations. Then one future observation was generated The experiment was repeated 100 times. In this case, the actual number of times the prediction interval contained the future observation was 78.

It is important to note that if only one baseline sample is taken and only one prediction interval for $k = 1$ future observation is computed, then the number of future observations out of a total of N future observations that will be contained in that one prediction interval is a binomial random variable with parameters $n = N$ and $p = 1-\alpha^*$, where α^* depends on the true population parameters and the computed bounds of the single prediction interval.

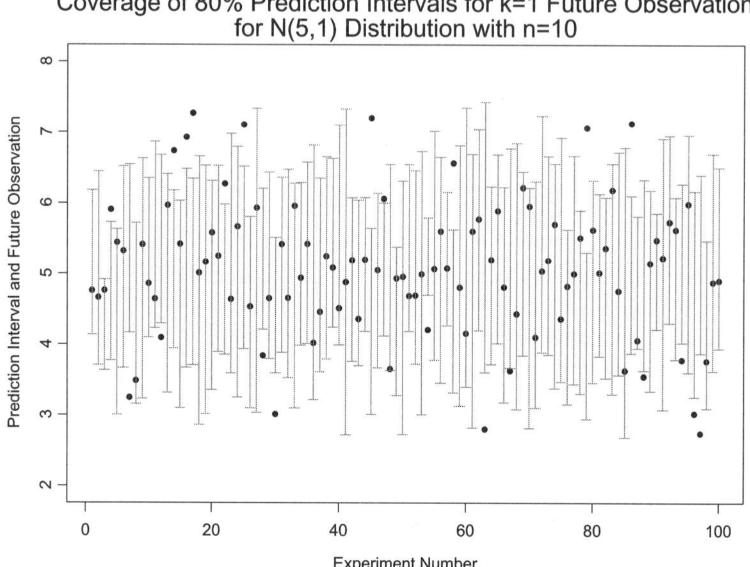

Figure 6.1. Results of simulation experiment showing the 80% prediction interval and one future observation for 100 simulations

6.2.1   Prediction Intervals for a Normal Distribution

Table 6.2 below shows arsenic concentrations (ppb) collected quarterly at two groundwater monitoring wells. These data are modified from benzene data given in USEPA (1992c, p. 56). In ENVIRONMENTALSTATS for S-PLUS these data are stored in the data frame `arsenic3.df`. In this example, we will use the data from the background well to construct a prediction interval for the next $k = 4$ observations.

Well	Year	Arsenic (ppb)			
Background	1	12.6	30.8	52.0	28.1
	2	33.3	44.0	3.0	12.8
	3	58.1	12.6	17.6	25.3
Compliance	4	48.0	30.3	42.5	15.0
	5	47.6	3.8	2.6	51.9

Table 6.2. Arsenic data from groundwater monitoring wells.

Combining all of the observations from the background well and assuming the data at the background well come from a normal distribution, the exact one-sided upper 95% prediction limit for the next $k = 4$ future observations is 72.9 ppb, and the one based on the Bonferroni method is 73.7. The four observed values of arsenic at the compliance well in year 4 are all below both of these prediction limits, as are all four values for year 5, so there is no evidence of contamination in either year.

Of course, this example ignores all sorts of possible design problems, including the fact that there is only one background well (so there is no way to estimate spatial variability). Also, sampling ceased at the background well in years 4 and 5 so there is no way to determine whether a change may have occurred in background concentrations during these years.

Menu

To construct a prediction interval for the arsenic concentrations using the ENVIRONMENTALSTATS for S-PLUS pull-down menu, follow these steps.

1. In the Object Explorer, find the data frame **arsenic3.df**.
2. On the S-PLUS menu bar, make the following menu choices: **EnvironmentalStats>Estimation>Prediction Intervals**. This will bring up the Prediction Interval dialog box.
3. For Data to Use select **Pre-Defined Data**. For Data Set select **arsenic3.df**. For Variable select **Arsenic**. In the Subset Rows with box, type **Well.type=="Background"**.
4. Click on the **Interval** tab. In the Distribution/Sample Size section, for Type select **Parametric**, for Distribution select **Normal**, and for n(mean) type **1**. Under the Prediction Interval section, **uncheck** the Simultaneous box, type **4** in the # Future Obs box, in the PI Type box select **upper**, in the PI Method box select **exact**, and for Confidence Level (%) select **95**.
5. Click **OK** or **Apply**.

These steps produce an exact upper prediction limit. To produce the upper prediction limit based on the Bonferroni method, in Step 4 above select **Bonferroni** in the PI Method box.

Command

To construct the exact prediction interval for the arsenic concentrations type these commands.

```
> attach(arsenic3.df)
> pred.int.norm(Arsenic[Well.type=="Background"], k=4,
    method="exact", pi.type="upper")
> detach()
```

To compute the prediction interval based on the Bonferroni method, replace `method="exact"` with `method="Bonferroni"`.

6.2.2   Prediction Intervals for a Lognormal Distribution

A prediction interval for a lognormal distribution is constructed by simply taking the natural logarithm of the observations and constructing a prediction interval based on the normal distribution, then exponentiating the prediction limits to produce a prediction interval on the original scale of the data (Hahn and Meeker, 1991, p. 73). In fact, you can use any monotonic transformation of the observations that you think induces normality (e.g., a Box-Cox power transformation), compute the prediction interval on the transformed scale, and then use the inverse transformation on the prediction limits to produce a prediction interval on the original scale.

To construct a prediction interval for a lognormal distribution using the ENVIRONMENTALSTATS for S-PLUS pull-down menu, follow steps similar to those shown in the previous section for a normal distribution, except choose **Lognormal** in the distribution box. To construct a prediction interval for a lognormal distribution using the ENVIRONMENTALSTATS for S-PLUS Command or Script Window, type commands similar to those shown in the previous section for a normal distribution, except use the function `pred.int.lnorm` instead of `pred.int.norm`.

We can use a prediction interval to compare the TcCB concentrations between the Cleanup and Reference areas (see Figures 1.1 and 1.2 on page 10). Based on the data from the Background area, the one-sided upper 95% prediction limit for the next $k = 77$ observations (there are 77 observations in the Cleanup area) is 2.68 ppb. There are 7 observations in the Cleanup area larger than 2.68, so the prediction interval indicates residual contamination is present in the Cleanup area. Note that both Student's t-test and the Wilcoxon rank sum test do not yield a significant difference between the two areas.

Menu

To produce the one-sided upper prediction limit based on the Reference area TcCB data, follow these steps.

1.   In the Object Explorer, highlight the data frame **epa.94b.tccb.df**.
2.   On the S-PLUS menu bar, make the following menu choices: **EnvironmentalStats>Estimation>Prediction Intervals**. This will bring up the Prediction Interval dialog box.
3.   For Data to Use select **Pre-Defined Data**. For Data Set select **epa.94b.tccb.df**. For Variable select **TcCB**. In the Subset Rows with box, type **Area=="Reference"**.
4.   Click on the **Interval** tab. In the Distribution/Sample Size section, for Type select **Parametric**, for Distribution select **Lognormal**, and for n(mean) type **1**. Under the Prediction Interval section, **uncheck** the Simultaneous box, type **77** in the # Future Obs box, in the PI Type box select **upper**, in the PI Method box select **exact**, and for Confidence Level (%) select **95**.
5.   Click **OK** or **Apply**.

Command

To produce the one-sided upper prediction limit for the next $k = 77$ observations based on the Reference area TcCB data, type these commands.

```
> attach(epa.94b.tccb.df)
> pred.int.lnorm(TcCB[Area=="Reference"], k=77,
    method="exact", pi.type="upper", conf.level=0.95)
> detach()
```

6.2.3 Nonparametric Prediction Intervals

You can construct prediction intervals without making any assumption about the distribution of the background data, except that the distribution is continuous. These kind of prediction intervals are called **nonparametric prediction intervals**. Of course, nonparametric prediction intervals still require the assumption that the distribution of future observations is the same as the distribution of the observations used to create the prediction interval.

Nonparametric prediction intervals are based on the ranked data. For example Table 6.3 illustrates the confidence levels associated with a one-sided upper prediction interval for the next $m = 3$ observations, based on various sample sizes, assuming the upper prediction limit is the maximum value. You can see that a confidence level greater than 95% cannot be achieved until the sample size is larger than $n = 50$. See Chapter 6 of Millard and Neerchal (2001) for a detailed discussion of nonparametric prediction intervals.

Sample Size (n)	Confidence Level (%)
5	63
10	77
15	83
20	87
25	89
50	94
75	96
100	97

Table 6.3. Confidence levels for one-sided upper nonparametric prediction interval for the next $k = 3$ observations, based on using the maximum value as the upper prediction limit.

The guidance document *Statistical Analysis of Ground-Water Monitoring Data at RCRA Facilities: Addendum to Interim Final Guidance* (USEPA, 1992c, pp. 59–60) gives an example of constructing a nonparametric prediction interval for the next $m = 2$ monthly observations of arsenic concentrations (ppb) in groundwater at a downgradient well, based on observations from three background wells. The data are shown in Table 6.4 and are stored in the data frame `epa.92c.arsenic2.df` in ENVIRONMENTALSTATS for S-PLUS.

Month	Background			Compliance
	Well 1	Well 2	Well 3	Well 4
1	<5	7	<5	
2	<5	6.5	<5	
3	8	<5	10.5	
4	<5	6	<5	
5	9	12	<5	8
6	10	<5	9	14

Table 6.4. Arsenic data (ppb) from groundwater monitoring wells.

The three background wells were sampled once per month for 6 months. The compliance well was only sampled in the 5th and 6th months. The EPA guidance document combines all of the observations from the three background wells ($n = 18$) and uses the maximum value 12 as an upper prediction limit for the next $m = 2$ observations at the compliance well. This produces a 90% upper prediction interval. Since one of the values from the compliance well lies above the upper prediction limit, we might conclude there is evidence of contamination at the compliance well, but we should keep in mind that given the way we constructed our prediction interval, we would incorrectly declare contamination present when in fact it is not present about 10% of the time.

Menu

To construct the nonparametric prediction interval using the ENVIRONMENTALSTATS for S-PLUS pull-down menu, follow these steps.

1. In the Object Explorer find the data frame **epa.92c.arsenic2.df**.
2. On the S-PLUS menu bar, make the following menu choices: **EnvironmentalStats>Estimation>Prediction Intervals**. This will bring up the Prediction Interval dialog box.
3. For Data to Use select **Pre-Defined Data**. For Data Set select **epa.92c.arsenic2.df**. For Variable select **Arsenic**. In the Subset Rows with box, type **Well.type=="Background"**.
4. Click on the **Interval** tab. In the Distribution/Sample Size section, set the Type button to **Nonparametric**. Under the Prediction Interval section, in the # Future Obs box type **2**, in the Min # Obs PI Should Contain box type **2**, in the PI Type box select **upper**, and in the Lower Bound box type **0**.
5. Click **OK** or **Apply**.

Command

To construct the nonparametric prediction interval type these commands.

```
> attach(epa.92c.arsenic2.df)
> pred.int.npar(Arsenic[Well.type=="Background"], m=2,
    lb=0, pi.type="upper")
> detach()
```

6.3 Simultaneous Prediction Intervals

Analyzing data from a groundwater monitoring program involves several difficulties, including trying to control for natural spatial and temporal variability, and perhaps dealing with nondetect values. One of the main statistical problems that plague groundwater monitoring programs at hazardous and solid waste facilities is the requirement of testing several wells and several constituents at each well on each sampling occasion. The number of constituents monitored can range from around 5 to 60 or more, and some facilities may have as many as 150 monitoring wells (Davis and McNichols, 1999). This is an obvious multiple comparisons problem, and the naïve approach of using a prediction interval with a conventional confidence level (e.g., 95% or 99%) for each comparison of a compliance well with background for each chemical of concern leads to a very high probability of at least one declaration of contamination on each sampling occasion, when in fact no contamination has occurred at any of the wells at any time for any of the chemicals of concern. This problem was pointed out several years ago by Millard (1987a) and others.

Davis and McNichols (1987, 1994b, 1999) proposed simultaneous prediction intervals as a way of controlling the facility-wide false positive rate (FWFPR) while maintaining adequate power to detect contamination in the groundwater. A ***simultaneous prediction interval*** is a prediction interval that will contain a certain number of future observations with probability $(1-\alpha)100\%$ for each of r future sampling occasions, where r is some pre-specified positive integer. The quantity r may refer to r distinct future sampling occasions in time at a single compliance well, or it may refer to sampling at r distinct compliance wells on one future sampling occasion. (Note: Current regulations prohibit evaluating false positive risks over more than one monitoring event, so for current practice r must refer to the number of compliance wells monitored, and "r future sampling occasions" refers to sampling each of the r wells once.) In either case, it is assumed that the distribution of concentrations is constant over all r future sampling occasions.

Although simultaneous prediction intervals help us control the Type I error rate (probability of declaring contamination when it is not present) over r monitoring wells, we need to control the Type I error rate over all monitoring wells *and* all constituents (chemicals and physical properties) we monitor. To do this, Davis and McNichols (1994b, 1999) suggest using the Bonferroni method and creating simultaneous prediction limits with confidence level $(1-\alpha/n_c)100\%$ for each of the n_c constituents being monitored. That is, for each constituent, the probability of mistakenly declaring contamination present at at least one of the compliance wells when in fact it is not present at any of the compliance wells (i.e., the significance level) is α/n_c.

There are several ways to define a rule for a simultaneous prediction limit. ENVIRONMENTALSTATS for S-PLUS includes a menu item and functions that allow for the following three rules:

- **The k-of-m Rule**. For the k-of-m rule, at least k of the next m future observations will fall in the prediction interval with probability $(1-\alpha)100\%$ on each of the r future sampling occasions. If observations are being taken sequentially, for a particular sampling occasion (or monitoring well), up to m observations may be taken, but once k of the observations fall within the prediction interval, sampling can stop. If $m - (k-1)$ observations fall outside the prediction interval, then contamination is declared to be present. For example, suppose we have $r = 5$ monitoring wells and we want to use the 1-of-3 rule (i.e., $k = 1$ and $m = 3$). Then for the ith monitoring well ($i = 1, 2, 3, 4, 5$), if the first observation is in the interval, we can stop. If the first observation is outside the interval, we have to wait a specified time (e.g., a few weeks), and take a second observation. If the second observation is in the interval, we can stop. If the second observation is outside the interval, then we have to wait a specified time and take a third observation. If the third observation is in the interval, we can stop. If the third observation is outside the interval, then contamination is declared to be present. (Note that in the case $k = m$ and $r = 1$, a simultaneous prediction interval reduces to the simple prediction interval we have already discussed.)

- **California Rule**. This is the rule currently required in the state of California. For the California rule, with probability $(1-\alpha)100\%$, for each of the r future sampling occasions, either the first observation will fall in the prediction interval, or else all of the next $m - 1$ observations will fall in the prediction interval. That is, if the first observation falls in the prediction interval then sampling can stop. Otherwise, up to $m - 1$ more observations must be taken (with a sufficient waiting time between sampling occasions). If any of these subsequent $m - 1$ observations falls outside the interval, we declare contamination is present.

- **Modified California Rule**. For the Modified California rule, with probability $(1-\alpha)100\%$, for each of the r future sampling occasions, either the first observation will fall in the prediction interval, or else at least 2 out of the next 3 observations will fall in the prediction interval. That is, if the first observation falls in the prediction interval then sampling can stop. Otherwise, up to 3 more observations must be taken (with a sufficient waiting time between sampling occasions). If any two of these next three observations fall into the interval then sampling can stop. Otherwise, contamination is declared to be present.

6.3.1 Simultaneous Prediction Intervals for a Normal Distribution

Using the background well arsenic data in Table 6.2 of Section 6.2.1, Figure 6.2 displays the value of the 95% upper simultaneous prediction limit as a function of the number of future sampling occasions (r) for the 1-of-3 rule, the California rule with $m = 3$, and the Modified California rule. This figure shows that in this case the upper prediction limit is always largest for the California rule and smallest for the 1-of-3 rule. For example, if you have 10 compliance wells you need to compare to the background arsenic level, you would set the upper prediction limit to about 48 ppb for the 1-of-3 rule, about 53 ppb for the Modified California rule, and about 63 ppb for the California rule. The probability of declaring contamination at at least one of these 10 compliance wells when in fact no well is contaminated is about 5%. The naïve 95% upper prediction limit for one future observation at a single well is 59.4 ppb. If you used this limit for all 10 compliance wells (but did not allow any resampling at any of the wells), then the probability of declaring contamination at at least one of these wells when in fact no well is contaminated is around $1-0.95^{10} = 40\%$ or larger.

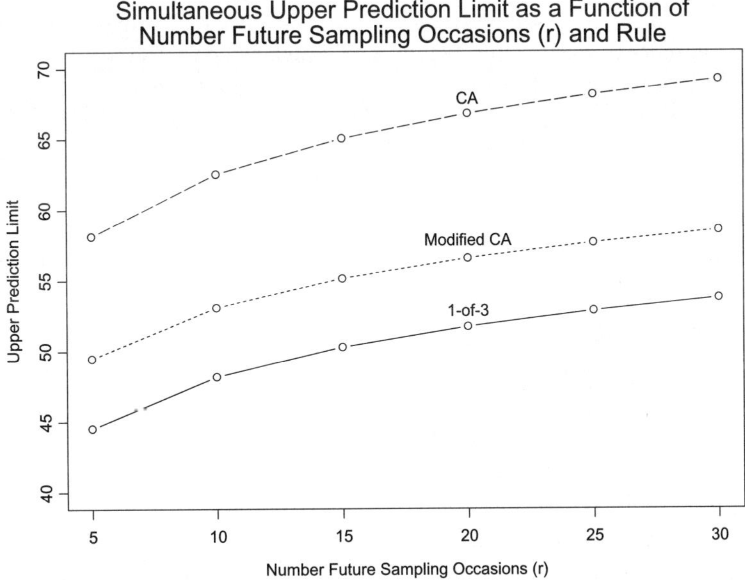

Figure 6.2. 95% upper prediction limit for arsenic as a function of the number of future sampling occasions (r) for the three different rules

Menu

To construct a simultaneous prediction interval for the arsenic concentrations using the ENVIRONMENTALSTATS for S-PLUS pull-down menu, follow these steps.

1. In the Object Explorer, find **arsenic3.df**.
2. On the S-PLUS menu bar, make the following menu choices: **EnvironmentalStats>Estimation>Prediction Intervals**. This will bring up the Prediction Interval dialog box.
3. For Data to Use select **Pre-Defined Data**. For Data Set select **arsenic3.df**. For Variable select **Arsenic**. In the Subset Rows with box, type **Well.type=="Background"**.
4. Click on the **Interval** tab. In the Distribution/Sample Size section, make sure the Type button is set to **Parametric** and that **Normal** is selected in the Distribution box. Under the Prediction Interval section, **check** the Simultaneous box, type **3** in the # Future Obs box, type **1** in the Min # Obs PI Should Contain box, type **10** in the # Future Sampling Occasions box, select **k.of.m** in the Rule box, in the PI Type box select **upper**, and in the Conf Level (%) box select **95**.
5. Click **OK** or **Apply**.

These steps produce a simultaneous 95% upper prediction limit for the 1-of-3 rule for $r = 10$ future sampling occasions. To use different values of r, type a different number in the # Future Sampling Occasions box in Step 4 above. To use the California rule, select **California** in the Rule box in Step 4 above, and to use the Modified California rule, select **Modified California** in the Rule box in Step 4 above.

Command

To construct a simultaneous prediction interval for the arsenic concentrations type these commands.

```
> attach(arsenic3.df)
> pred.int.norm.simultaneous(
    Arsenic[Well.type=="Background"], k=1, m=3, r=10,
    rule="k.of.m", pi.type="upper", conf.level=0.95)
```

These steps produce a simultaneous 95% upper prediction limit for the 1-of-3 rule for $r = 10$ future sampling occasions. To use different values of r, set the argument r to a different number.

To use the California and Modified California rules, type these commands.

```
> pred.int.norm.simultaneous(
    Arsenic[Well.type=="Background"], m=3, r=10,
    rule="CA", pi.type="upper", conf.level=0.95)
> pred.int.norm.simultaneous(
    Arsenic[Well.type=="Background"], r=10,
    rule="Modified.CA", pi.type="upper", conf.level=0.95)
```

To detach the `arsenic3.df` data frame from your search list, type this command.

```
> detach()
```

6.3.2 Simultaneous Prediction Intervals for a Lognormal Distribution

Just as for a standard prediction interval for a lognormal distribution, a simultaneous prediction interval for a lognormal distribution is constructed by simply taking the natural logarithm of the observations and constructing a simultaneous prediction interval based on the normal distribution, then exponentiating the prediction limits to produce a simultaneous prediction interval on the original scale of the data. To construct a simultaneous prediction interval for a lognormal distribution using the ENVIRONMENTALSTATS for S-PLUS pull-down menu, follow steps similar to those shown in the previous section for a normal distribution, except choose **Lognormal** in the Distribution box. To construct a simultaneous prediction interval for a lognormal distribution using the ENVIRONMENTALSTATS for S-PLUS Command or Script Window, type commands similar to those shown in the previous section, except use the function `pred.int.lnorm.simultaneous`.

6.3.3 Simultaneous Nonparametric Prediction Intervals

Chou and Owen (1986) developed the theory for nonparametric simultaneous prediction limits for various rules, including the 1-of-m rule. Their theory, however, does not cover the California or Modified California rules, and uses an r-fold summation involving a minimum of 2^r terms. Davis and McNichols (1994b; 1999) extended the results of Chou and Owen (1986) to include the California and Modified California rule, and developed algorithms that involve summing far fewer terms.

Like a standard nonparametric prediction interval, a simultaneous nonparametric prediction interval is based on the order statistics from the sample. For a one-sided upper simultaneous nonparametric prediction interval, the upper prediction limit is usually the largest observation in the background data, but it could be the next largest or any other order statistic. Similarly, for a one-sided lower simultaneous nonparametric prediction interval, the lower prediction limit is usually the smallest observation.

In Section 6.2.3 we showed how to create a one-sided upper nonparametric prediction interval for the next two observations of arsenic at a single downgradient well based on $n = 18$ observations at three upgradient wells (see Table 6.4). In that example, the significance level associated with this prediction interval is 10%.

Now suppose there are $r = 5$ downgradient wells, nine other constituents are being monitored besides arsenic (i.e., $n_c = 10$), and we want a FWFPR of 5% for each sampling occasion. Then the significance level for the simultaneous prediction interval for arsenic should be set to $0.05/10 = 0.005$, which means the confidence level should be set to $(1-0.005)100\% = 99.5\%$. Using the maximum value of 12 ppb as the upper prediction limit, Table 6.5 displays the associated confidence levels for various plans. The 1-of-3 and 1-of-4 plans satisfy the required confidence level, but neither of the California plans nor the Modified California plan do. Note that for k-of-m plans, the confidence level increases with increasing values of m, while the opposite is true for California plans. In order for a California plan with $m = 3$ to satisfy the required confidence level of 99.5%, a total of at least $n = 61$ background observations are needed. In order for the Modified California plan to satisfy the required confidence level, a total of at least $n = 24$ background observations are needed.

Rule	k	m	Confidence Level (%)
k-of-m	1	3	99.6
	1	4	99.9
CA		3	95.5
		4	93.9
Modified CA			99.1

Table 6.5. Confidence levels for various nonparametric simultaneous prediction intervals based on $n = 18$ background observations and $r = 5$ monitoring wells.

Menu

To construct the simultaneous nonparametric prediction intervals and obtain their confidence levels (shown in Table 6.5) follow these steps.

1. In the Object Explorer, highlight the data frame **epa.92c.arsenic2.df**.
2. On the S-PLUS menu bar, make the following menu choices: **EnvironmentalStats>Estimation>Prediction Intervals**. This will bring up the Prediction Interval dialog box.
3. For Data to Use select **Pre-Defined Data**. For Data Set select **epa.92c.arsenic2.df**. For Variable select **Arsenic**. In the Subset Rows with box, type **Well.type=="Background"**.
4. Click on the **Interval** tab. In the Distribution/Sample Size section, set the Type button to **Nonparametric**.
5. Under the Prediction Interval section, **check** the Simultaneous box, in the # Future Obs box type **3**, in the Min # Obs PI Should Contain box type **1**, and in the # Future Sample Occasions box type **5**. In the Rule box select **k of m**, in the PI Type box select **upper**, and in the Lower Bound box type **0**.
6. Click **OK** or **Apply**.

The above steps produce information for the 1-of-3 plan. To produce information for the 1-of-4 plan, Step 5 above becomes:

5. Under the Prediction Interval section, **check** the Simultaneous box, in the # Future Obs box type **4**, in the Min # Obs PI Should Contain box type **1**, and in the # Future Sample Occasions box type **5**. In the Rule box select **k of m**, in the PI Type box select **upper**, and in the Lower Bound box type **0**.

To produce information for the California plan with $m = 3$, Step 5 above becomes:

5. Under the Prediction Interval section, **check** the Simultaneous box, in the # Future Obs box type **3** and in the # Future Sample Occasions box type **5**. In the Rule box select **CA**, in the PI Type box select **upper**, and in the Lower Bound box type **0**.

To produce information for the California plan with $m = 4$, Step 5 above becomes:

5. Under the Prediction Interval section, **check** the Simultaneous box, in the # Future Obs box type **4** and in the # Future Sample Occasions box type **5**. In the Rule box select **CA**, in the PI Type box select **upper**, and in the Lower Bound box type **0**.

Finally, to produce information for the Modified California plan, Step 5 above becomes:

5. Under the Prediction Interval section, **check** the Simultaneous box, and in the # Future Sample Occasions box type **5**. In the Rule box select **Modified CA**, in the PI Type box select **upper**, and in the Lower Bound box type **0**.

Command

To construct the simultaneous nonparametric prediction intervals and obtain their confidence levels (shown in Table 6.5), type these commands.

```
> attach(epa.92c.arsenic2.df)
> pred.list.1.of.3 <- pred.int.npar.simultaneous(
    Arsenic[Well.type=="Background"], k=1, m=3, r=5,
    rule="k.of.m", lb=0, pi.type="upper")
> print(pred.list.1.of.3, conf.cov.sig.digits=7)
> pred.list.1.of.4 <- pred.int.npar.simultaneous(
    Arsenic[Well.type=="Background"], k=1, m=4, r=5,
    rule="k.of.m", lb=0, pi.type="upper")
> print(pred.list.1.of.4, conf.cov.sig.digits=7)
> pred.list.CA.w.3 <- pred.int.npar.simultaneous(
    Arsenic[Well.type=="Background"], m=3, r=5,
    rule="CA", lb=0, pi.type="upper")
> print(pred.list.CA.w.3, conf.cov.sig.digits=7)
```

```
> pred.list.CA.w.4 <- pred.int.npar.simultaneous(
    Arsenic[Well.type=="Background"], m=4, r=5,
    rule="CA", lb=0, pi.type="upper")
> print(pred.list.CA.w.4, conf.cov.sig.digits=7)
> pred.list.Modified.CA <- pred.int.npar.simultaneous(
    Arsenic[Well.type=="Background"], r=5,
    rule="Modified.CA", lb=0, pi.type="upper")
> print(pred.list.Modified.CA, conf.cov.sig.digits=7)
> detach()
```

6.4   Tolerance Intervals

A *tolerance interval* for some population is an interval on the real line constructed so as to contain $\beta100\%$ of the population (i.e., $\beta100\%$ of all future observations), where $0 < \beta < 1$ (usually β is bigger than 0.5). The quantity $\beta100\%$ is called the *coverage*. (Note: Do not confuse our use of the symbol β here with the probability of a Type II error. The symbol β is used here to be consistent with previous literature on tolerance intervals.)

Table 6.6 lists the functions available in ENVIRONMENTALSTATS for S-PLUS for constructing tolerance intervals. From the S-PLUS menu bar, you can construct prediction intervals by making the selection **Environmental-Stats>Estimation>Tolerance Intervals**.

Distribution	Functions for Prediction Intervals	Description
Normal	`tol.int.norm`	Construct a tolerance interval for a normal distribution.
	`tol.int.norm.K`	Compute the value of K for a tolerance interval for a normal distribution
Lognormal	`tol.int.lnorm`	Construct a tolerance interval for a lognormal distribution.
Poisson	`tol.int.pois`	Construct a tolerance interval for a Poisson distribution.
Nonparametric	`tol.int.npar`	Construct a nonparametric tolerance interval.

Table 6.6. Functions in ENVIRONMENTALSTATS for S-PLUS for constructing tolerance intervals.

As with a prediction interval, the basic idea of a tolerance interval is to assume a particular probability distribution (e.g., normal, lognormal, etc.) for some process generating the data (e.g., quarterly observations of chemical concentrations in groundwater), compute sample statistics from a baseline sample, and then use these sample statistics to construct a tolerance interval, assuming the distribution of the data does not change in the future. For example, if X denotes a random variable from some population, and we know what the population looks like (e.g., N(10, 2)) so we can compute the quantiles of the population, then a $\beta 100\%$ two-sided tolerance interval is given by:

$$\left[\, x_{1-\beta/2} \, , \; x_{\beta/2} \,\right] \tag{6.4}$$

where x_p denotes the pth quantile of the distribution of X. Similarly, a $\beta 100\%$ one-sided upper tolerance interval is given by:

$$\left[\, -\infty \, , \; x_\beta \,\right] \tag{6.5}$$

and a $\beta 100\%$ one-sided lower tolerance interval is given by:

$$\tag{6.6}$$

Note that in the case when the distribution of X is known, a $\beta 100\%$ tolerance interval is exactly the same as a $(1-\alpha)100\%$ prediction interval for $k = 1$ future observation, where $\beta = 1-\alpha$ (see Equations (6.1) to (6.3)).

Usually the true distribution of X is unknown, so the values of the tolerance limits have to be estimated based on estimating the parameters of the distribution of X. In this case, a tolerance interval is a random interval; that is, the lower and/or upper bounds are random variables computed based on sample statistics in the baseline sample. Given this uncertainty in the bounds, there are two ways to construct tolerance intervals (Guttman, 1970):

- A **β-content** tolerance interval with **confidence level** $(1-\alpha)100\%$ is constructed so that it contains *at least* $\beta 100\%$ of the population (i.e., the coverage is at least $\beta 100\%$) with probability $(1-\alpha)100\%$.
- A **β-expectation** tolerance interval is constructed so that it contains *on average* $\beta 100\%$ of the population (i.e., the average coverage is $\beta 100\%$).

A β-expectation tolerance interval with coverage $\beta 100\%$ is equivalent to a prediction interval for $k = 1$ future observation with associated confidence level $\beta 100\%$. Note that there is no explicit confidence level associated with a β-expectation tolerance interval. If a β-expectation tolerance interval is treated as a β-content tolerance interval, the confidence level associated with this tolerance interval is usually around 50% (e.g., Guttman, 1970, Table 4.2, p. 76). Thus, a β-content tolerance interval with coverage $\beta 100\%$ will usually be wider than a β-expectation tolerance interval with the same coverage if the confidence level associated with the β-content tolerance interval is more than 50%.

It can be shown (e.g., Conover, 1980, pp. 119–121) that an upper confidence interval for the pth quantile with confidence level $(1-\alpha)100\%$ is equivalent to an upper β-content tolerance interval with coverage $100p\%$ and confidence level $(1-\alpha)100\%$. Also, a lower confidence interval for the pth quantile with confidence level $(1-\alpha)100\%$ is equivalent to a lower β-content tolerance interval with coverage $100(1-p)\%$ and confidence level $(1-\alpha)100\%$.

Prediction and tolerance intervals have long been applied to quality control and life testing problems. In environmental monitoring, USEPA has proposed using tolerance intervals in at least two different ways: compliance-to-background comparisons and compliance-to-fixed standard comparisons. See Chapter 6 of Millard and Neerchal (2001) for a discussion of this topic.

6.4.1   Tolerance Intervals for a Normal Distribution

Section 5.3.1 on page 100 discusses a permit in which an ACL of 50 ppm for aldicarb should not be exceeded more than 5% of the time. Using the data from three groundwater monitoring compliance wells (four monthly samples at each well) stored in `epa.89b.aldicarb2.df`, we computed a lower 99% confidence limit for the 95th percentile for the distribution at each of the three compliance wells, yielding 25.3, 25.7, and 25.5 ppm. This is equivalent to computing a lower β-content tolerance limit with coverage 5% and associated confidence level of 99%.

Menu

To produce the one-sided lower β-content tolerance limits with coverage 5% and associated confidence level 99% for each of the compliance wells using the ENVIRONMENTALSTATS for S-PLUS pull-down menu, follow these steps.

1.   In the Object Explorer, find **epa.89b.aldicarb2.df**.
2.   On the S-PLUS menu bar, make the following menu choices: **EnvironmentalStats>Estimation>Tolerance Intervals**. This will bring up the Tolerance Interval dialog box.
3.   For Data to Use select **Pre-Defined Data**. For Data Set box select **epa.89b.aldicarb2.df**. For Variable select **Aldicarb**. In the Subset Rows with box, type **Well=="1"**.
4.   Click on the **Interval** tab. Under the Distribution group, for Type, make sure the **Parametric** button is selected, and in the Distribution box select **Normal**. Under the Tolerance Interval group, for Coverage Type select the **Content** button, for TI Type select **lower**, in the Coverage (%) box select **5**, and in the Confidence Level (%) box select **99**.
5.   Click **OK** or **Apply**.

This will produce the one-sided lower β-content tolerance limit for the first well. To produce the tolerance limits for the other two wells, in Step 3 above, in the Subset Rows with box, type **Well=="2"** or **Well=="3"**.

Command

To produce the one-sided lower β-content tolerance limits with coverage 5% and associated confidence level 99% for each of the compliance wells type these commands.

```
> attach(epa.89b.aldicarb2.df)
```

```
> tol.int.norm(Aldicarb[Well=="1"], coverage=0.05,
    cov.type="content", ti.type="lower", conf.level=0.99)
> tol.int.norm(Aldicarb[Well=="2"], coverage=0.05,
    cov.type="content", ti.type="lower", conf.level=0.99)
> tol.int.norm(Aldicarb[Well=="3"], coverage=0.05,
    cov.type="content", ti.type="lower", conf.level=0.99)
> detach()
```

Instead of calling `tol.int.norm` three separate times, you could instead just use the following single command:

```
> lapply(split(Aldicarb, Well), tol.int.norm,
    coverage=0.05, cov.type="content", ti.type="lower",
    conf.level=0.99)
```

6.4.2   Tolerance Intervals for a Lognormal Distribution

In Section 6.2.2 we computed a prediction interval to compare the TcCB concentrations between the Cleanup and Reference areas (see Figures 1.1 and 1.2 on page 10). Based on the data from the Background area, the one-sided upper 95% prediction limit for the next $k = 77$ observations (there are 77 observations in the Cleanup area) is 2.68 ppb. There are 7 observations in the Cleanup area larger than 2.68, so the prediction interval indicates residual contamination is present in the Cleanup area.

Some guidance documents suggest constructing a one-sided upper tolerance interval based on the Reference area and comparing all of the observations from the Cleanup area to the upper tolerance limit. This is sometimes called the "Hot-Measurement Comparison" (USEPA, 1994b, pp. 4.8–4.9). Chapter 6 of Millard and Neerchal (2001) explains why this method should never be used because you do not know the true Type I error rate. In this case, the one-sided upper 95% β-content tolerance limit with associated confidence level 95% is 1.42 ppb (versus 2.68 ppb for the upper prediction limit).

Menu

To produce the one-sided upper β-content tolerance limit based on the Reference area TcCB data follow these steps.

1. In the Object Explorer, highlight **epa.94b.tccb.df**.
2. On the S-PLUS menu bar, make the following menu choices: **EnvironmentalStats>Estimation>Tolerance Intervals**. This will bring up the Tolerance Interval dialog box.
3. For Data to Use select **Pre-Defined Data**. For Data Set select **epa.94b.tccb.df**. For Variable select **TcCB**. In the Subset Rows with box, type **Area=="Reference"**.

4. Click on the **Interval** tab. Under the Distribution group, for Type, make sure the **Parametric** button is selected, and in the Distribution box select **Lognormal**. Under the Tolerance Interval group, for Coverage Type select the **Content** button, for TI Type select **upper**, in the Coverage (%) box select **95**, and in the Confidence Level (%) box select **95**.

5. Click **OK** or **Apply**.

Command

To produce the one-sided upper β-content tolerance limit type these commands.

```
> attach(epa.94b.tccb.df)
> tol.int.lnorm(TcCB[Area=="Reference"], coverage=0.95,
    cov.type="content", ti.type="upper", conf.level=0.95)
> detach()
```

6.4.3   Nonparametric Tolerance Intervals

You can construct tolerance intervals without making any assumption about the distribution of the background data, except that the distribution is continuous. These kind of tolerance intervals are called ***nonparametric tolerance intervals***. Of course, nonparametric tolerance intervals still require the assumption that the distribution of future observations is the same as the distribution of the observations used to create the tolerance interval. Just as for nonparametric prediction intervals, nonparametric tolerance intervals are based on the ranked data.

In Section 5.3.3 on page 104 we created a one-sided upper confidence interval for the 95th percentile of copper concentrations based on $n = 24$ observations from three background wells. The upper limit was the maximum value of 9.2 ppb and the confidence interval for the 95th percentile had an associated confidence level of 71%. The upper limit was constructed with the idea that this limit will be used as a threshold value for concentrations observed in two compliance wells (i.e., if any concentrations at the compliance wells exceed this limit, this indicates there may be contamination in the groundwater). This is the "Hot-Measurement Comparison," and as already discussed above there are some major problems with this technique. Creating a one-sided upper confidence interval for the 95th percentile is equivalent to creating a one-sided upper β-content tolerance interval with 95% coverage. Millard and Neerchal (2001, p. 349) compare this method with using a simultaneous prediction limit.

Menu

To produce the nonparametric one-sided upper β-content tolerance interval with coverage 95% and associated confidence level 71% for the copper concentrations using the ENVIRONMENTALSTATS for S-PLUS pull-down menu, follow these steps.

1. In the Object Explorer, highlight **epa.92c.copper2.df**.
2. On the S-PLUS menu bar, make the following menu choices: **EnvironmentalStats>Estimation>Tolerance Intervals**.   This will bring up the Tolerance Interval dialog box.
3. For Data to Use select **Pre-Defined Data**. For Data Set select **epa.92c.copper2.df**. For Variable select **Copper**. In the Subset Rows with box, type **Well.type=="Background"**.
4. Click on the **Interval** tab. Under the Distribution group, for Type, select the **Nonparametric** button. Under the Tolerance Interval group, for Coverage Type select the **Content** button, for Supply select the **Coverage** button, for TI Type select **upper**, in the Coverage (%) box select **95**, and in the Lower Bound box type **0**.
5. Click **OK** or **Apply**.

Command

To produce the nonparametric one-sided upper β-content tolerance interval with coverage 95% and associated confidence level 71% for the copper concentrations, type these commands.

```
> attach(epa.92c.copper2.df)
> tol.int.npar(Copper[Well.type=="Background"],
    coverage=0.95, cov.type="content", ti.type="upper",
    lb=0)
> detach()
```

6.5 Summary

- Any activity that requires comparing new values to "background" or "standard" values creates a decision problem: if the new values greatly exceed the background or standard value, is this evidence of a true difference (i.e., is there contamination)?
- Statistical tests are used as objective tools to decide whether a change has occurred. For a monitoring program that involves numerous tests over time, figuring out how to balance the *overall* Type I error with the power of detecting a change is a multiple comparisons problem.
- Prediction intervals and tolerance intervals are two tools that you can use to attempt to solve the multiple comparisons problem.
- Table 6.1 lists the functions available in ENVIRONMENTALSTATS for S-PLUS for constructing prediction intervals. From the S-PLUS menu bar, you can construct prediction intervals by making the selection **EnvironmentalStats>Estimation>Prediction Intervals**.

- Table 6.6 lists the functions available in ENVIRONMENTALSTATS for S-PLUS for constructing tolerance intervals. From the S-PLUS menu bar, you can construct prediction intervals by making the selection **EnvironmentalStats>Estimation>Tolerance Intervals**.

7

Hypothesis Tests

7.1 Introduction

If you are comparing chemical concentrations between a "background" area and a "potentially contaminated" area, how different do the concentrations in these two areas have to be before you decide that the "potentially contaminated" area is in fact contaminated? In the last chapter we showed how to use prediction and tolerance intervals to try to answer this question. There are other kinds of hypothesis tests you can use as well. S-PLUS contains several menu items and functions for performing classical statistical hypothesis tests, such as t-tests, analysis of variance, linear regression, nonparametric tests, quality control procedures and time series analysis (see the S-PLUS documentation and help files). ENVIRONMENTALSTATS for S-PLUS contains menu items and functions for some statistical tests that are not included in S-PLUS but that are often used in environmental statistics, such as the Shapiro-Francia, Shapiro-Wilk, and Probability Plot Correlation Coefficient goodness-of-fit tests; Kendall's seasonal test for trend; and the quantile test for a shift in the tail of the distribution. This chapter discusses the menu items and functions available in ENVIRONMENTALSTATS for S-PLUS for these additional hypothesis tests. See Chapters 7, 9, and 11 of Millard and Neerchal (2001) for a more in-depth discussion of hypothesis tests.

7.2 Goodness-of-Fit Tests

Most commonly used parametric statistical tests assume the observations in the random sample come from a normal population. So how do you know whether this assumption is valid? We saw in Chapter 3 how to make a visual assessment of this assumption using Q-Q plots. Another way to verify this assumption is with a goodness-of-fit test, which lets you specify what kind of distribution you think the data come from and then compute a test statistic and a p-value.

A goodness-of-fit test may be used to test the null hypothesis that the data come from a specific distribution, such as "the data come from a normal distribution with mean 10 and standard deviation 2," or to test the more general null hypothesis that the data come from a particular family of distributions, such as "the data come from a lognormal distribution." Goodness-of-fit tests are mostly used to test the latter kind of hypothesis, since in practice we rarely know or want to specify the parameters of the distribution.

In practice, goodness-of-fit tests may be of limited use for very large or very small sample sizes. Almost any goodness-of-fit test will reject the null hypothesis of the specified distribution if the number of observations is very large, since "real" data are never distributed according to any theoretical distribution (Conover, 1980, p. 367). On the other hand, with only a very small number of observations, no test will be able to determine whether the observations appear to come from the hypothesized distribution or some other totally different looking distribution.

Table 7.1 lists the menu items and functions available in ENVIRONMENTALSTATS for S-PLUS for performing goodness-of-fit tests. All menu items fall under the menu choice **EnvironmentalStats>Hypothesis Tests>GOF Tests**. For example, the first entry in the table describes what you can do if you make the selection **EnvironmentalStats>Hypothesis Tests>GOF Tests>One Sample>Shapiro-Wilk**.

Menu Item	Functions	Description
One Sample> Shapiro-Wilk	`sw.gof`	Shapiro-Wilk goodness of fit test for normality.
One Sample> Shapiro-Francia	`sf.gof`	Shapiro-Francia goodness of fit test for normality.
One Sample> PPCC	`ppcc.norm.gof`	Probability plot correlation coefficient (PPCC) goodness of fit test for normality.
One Sample> PPCC (EVD)	`ppcc.evd.gof`	Probability plot correlation coefficient (PPCC) goodness of fit test for an extreme value distribution.
One Sample> Kolmogorov-Smirnov	`ks.gof`	Kolmogorov-Smirnov goodness-of-fit test to compare a sample with a specified probability distribution.
Two Samples> Kolmogorov-Smirnov	`ks.gof`	Kolmogorov-Smirnov goodness-of-fit test to compare two samples.
Group> Shapiro-Wilk	`sw.group.gof`	Shapiro-Wilk group goodness of fit test for normality.
Group> Shapiro-Francia	`sf.group.gof`	Shapiro-Francia group goodness of fit test for normality.
Group> PPCC	`ppcc.norm.group.gof`	Probability plot correlation coefficient (PPCC) group goodness of fit test for normality.

Table 7.1. Menu items and functions in ENVIRONMENTALSTATS for S-PLUS for goodness-of-fit tests.

7.2.1   One Sample Goodness-of-Fit Tests for Normality

The first three rows of Table 7.1 list the menu items and functions available for performing a one-sample goodness-of-fit test for normality. You can also use these menu items and functions to determine whether a set of observations appear to come from a lognormal, three-parameter lognormal, zero-modified normal, or zero-modified lognormal (delta) distribution.

In Chapters 1 and 3 we saw that the Reference area TcCB data appear to come from a lognormal distribution based on a histogram (Figure 1.1), an empirical cdf plot (Figure 1.4), a normal Q-Q plot (Figures 1.6), a Tukey mean-difference Q-Q plot (Figure 1.7), and a plot of the PPCC versus λ for a variety of Box-Cox transformations (Figure 3.7). Here we will formally test whether the Reference area TcCB data appear to come from a normal or lognormal distribution.

Table 7.2 lists the results of these two tests. The second and third columns show the test statistics with the p-values in parentheses. The p-values clearly indicate that we should not assume the Reference area TcCB data come from a normal distribution, but the assumption of a lognormal distribution appears to be adequate. Figure 7.1 and Figure 7.2 show companion plots for the results of the Shapiro-Wilk tests for normality and lognormality, respectively. These plots include the observed distribution overlaid with the fitted distribution, the observed and fitted cdf, the normal Q-Q plot, and the results of the hypothesis test.

Assumed Distribution	Shapiro-Wilk (W)	Shapiro-Francia (W')
Normal	0.918 (p=0.003)	0.923 (p=0.006)
Lognormal	0.979 (p=0.55)	0.987 (p=0.78)

Table 7.2. Results of tests for normality and lognormality for the Reference area TcCB data.

Menu

To perform the Shapiro-Wilk test for normality using the ENVIRONMENTALSTATS for S-PLUS pull-down menu, follow these steps.

1. In the Object Explorer, find and highlight the data frame **epa.94b.tccb.df** or **new.epa.94b.tccb.df**.

2. On the S-PLUS menu bar, make the following menu choices: **EnvironmentalStats>Hypothesis Tests>GOF Tests>One Sample>Shapiro-Wilk**. This will bring up the Shapiro-Wilk GOF Test dialog box.

3. For Data to Use, select **Pre-Defined Data**. For Data Set make sure **epa.94b.tccb.df** is selected, for Variable select **TcCB**, and in the Subset Rows with box type **Area=="Reference"**. In the Distribution box select **Normal**.

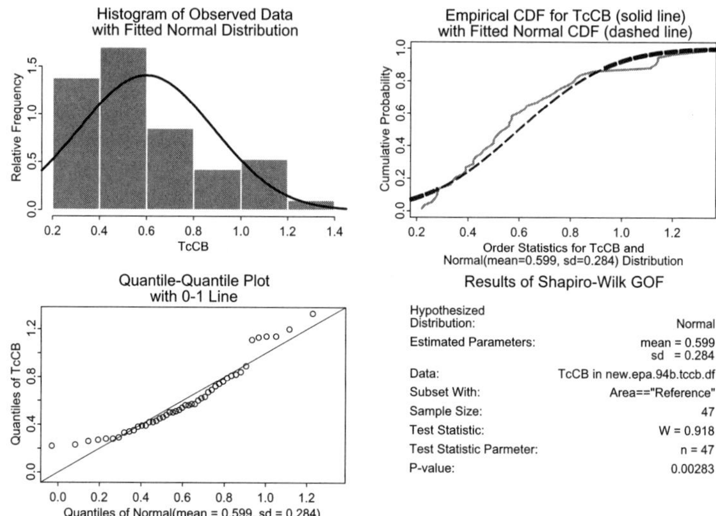

Figure 7.1. Companion plots for the Shapiro-Wilk test for normality for the Reference area TcCB data.

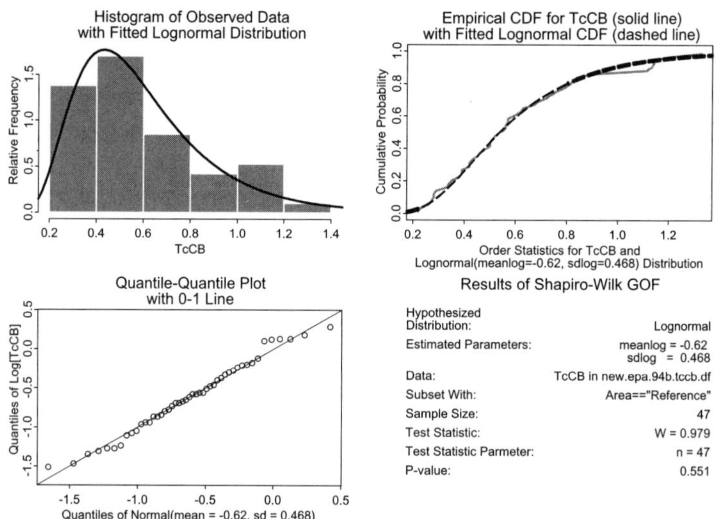

Figure 7.2. Companion plots for the Shapiro-Wilk test for lognormality for the Reference area TcCB data.

4. Click on the **Plotting** tab. Under Plotting Information, in the Significant Digits box select **3**, then click **OK** or **Apply**.

To perform the Shapiro-Wilk test for lognormality, repeat the above steps, but in Step 3 select **Lognormal** in the Distribution box. To perform the Shapiro-Francia tests, repeat the above procedure, but in Step 2 make the menu selection **EnvironmentalStats>Hypothesis Tests>GOF Tests>One Sample>Shapiro-Francia**.

Command

To perform the Shapiro-Wilk tests type these commands.

```
> attach(epa.94b.tccb.df)
> TcCB.ref <- TcCB[Area=="Reference"]
> sw.list.norm <- sw.gof(TcCB.ref)
> sw.list.norm
> sw.list.lnorm <- sw.gof(TcCB.ref, dist="lnorm")
> sw.list.lnorm
> detach()
```

To plot the results of these tests as shown in Figure 7.1 and Figure 7.2, type these commands.

```
> plot.gof.summary(sw.list.norm, digits=3)
> plot.gof.summary(sw.list.lnorm, digits=3)
```

To perform the Shapiro-Francia tests, type these commands.

```
> sf.list.norm <- sf.gof(TcCB.ref)
> sf.list.norm
> sf.list.lnorm <- sf.gof(TcCB.ref, dist="lnorm")
> sf.list.lnorm
```

To plot the results of the Shapiro-Francia tests, type these commands.

```
> plot.gof.summary(sf.list.norm, digits=3)
> plot.gof.summary(sf.list.lnorm, digits=3)
```

Note: You can also plot the results of goodness-of-fit tests by using the generic `plot` function. For example, typing `plot(sw.list.norm)` will bring up a menu of options for different kinds of plots, including the summary plots.

7.2.2   Testing Several Groups for Normality

If you have several sets of observations you want to test for normality, you may encounter the multiple comparisons problem. For example, current regulations for monitoring groundwater at hazardous and solid waste sites usually require performing statistical analyses even when there are only small sample sizes at each monitoring well. As we noted above, goodness-of-fit tests are not very useful with small sample sizes; there is simply not enough information to determine whether the data appear to come from the hypothesized distribution or not. Gibbons (1994, p. 228) suggests pooling the measures from several upgradient wells to establish "background." Due to spatial variability, the wells may have different means and variances, yet you would like to test the assumption of a normal distribution for the chemical concentration at each of the upgradient wells.

Wilk and Shapiro (1968) suggest two different test statistics for the problem of testing the normality of K separate groups, using the results of the Shapiro-Wilk test applied to random samples from each of the K groups. Both test statistics are functions of the K p-values that result from performing the test on each of the K samples. Under the null hypothesis that all K samples come from normal distributions, the p-values represent a random sample from a uniform distribution on the interval [0,1]. Since these two test statistics are based solely on the p-values, they are really meta-analysis statistics (Fisher and van Belle, 1993, p. 893), and can be applied to the problem of combining the results from K independent hypothesis tests, where the hypothesis tests are not necessarily goodness-of-fit tests. The last three rows of Table 7.1 list the menu items and functions available in ENVIRONMENTALSTATS for S-PLUS for performing group tests for normality.

The guidance document USEPA (1992c, p. 21) contains observations of arsenic concentrations (ppm) collected over 4 months at six monitoring wells. These data are displayed in Table 7.3 and are stored in the data frame epa.92c.arsenic1.df in ENVIRONMENTALSTATS for S-PLUS.

Month	Well 1	Well 2	Well 3	Well 4	Well 5	Well 6
1	22.9	2.0	2.0	7.84	24.9	0.34
2	3.09	1.25	109.4	9.3	1.3	4.78
3	35.7	7.8	4.5	25.9	0.75	2.85
4	4.18	52	2.5	2.0	27	1.2

Table 7.3. Arsenic concentrations (ppm) in groundwater.

Figure 7.3 displays the observations for each well, and Figure 7.4 displays the log-transformed observations. Here we will use the Shapiro-Wilk group test to test the null hypotheses that the observations at each well represent a sample from some kind of normal or lognormal distribution, but the population means and/or variances may differ between wells.

Table 7.4 displays the results of the normal scores and chi-squared scores tests for the null hypotheses of normality and lognormality. Figure 7.5 and Figure 7.6 display the normal Q-Q plots of the normal scores for each of the hypotheses. There is clear evidence that the concentrations do not come from normal distributions, but the assumption of lognormal distributions appears to be adequate.

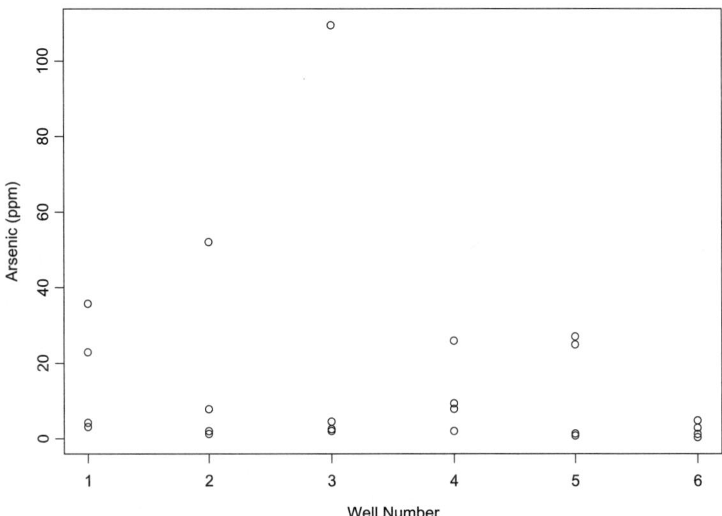

Figure 7.3. Arsenic concentrations (ppm) by well.

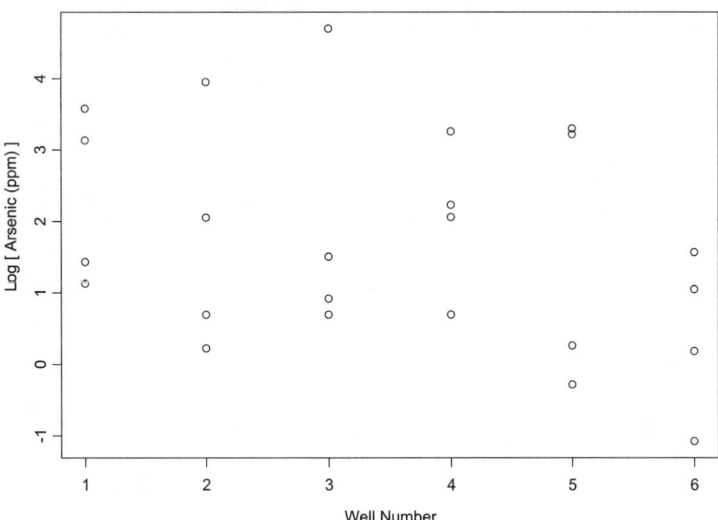

Figure 7.4. Log-transformed arsenic concentrations by well.

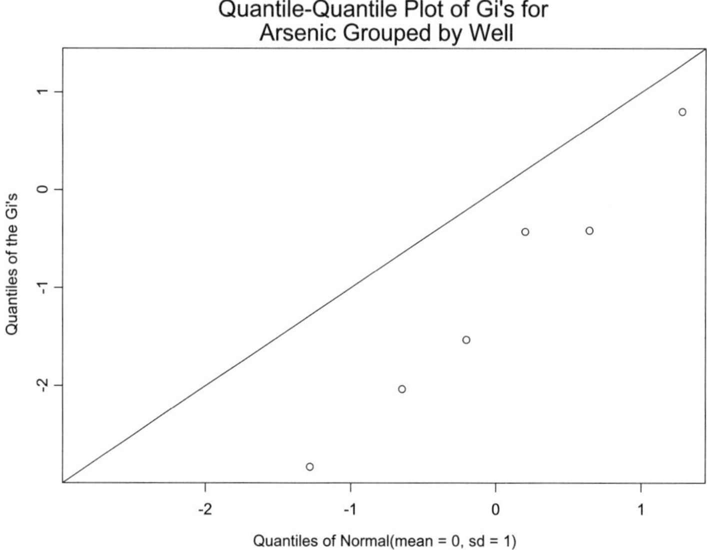

Figure 7.5. Normal Q-Q plot of normality scores used in the group Shapiro-Wilk goodness-of-fit test for normality of the arsenic concentrations.

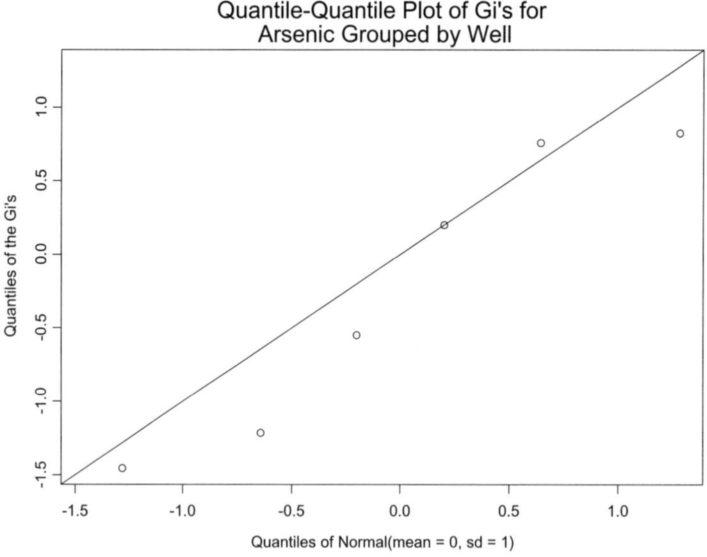

Figure 7.6. Normal Q-Q plot of normality scores used in the group Shapiro-Wilk goodness-of-fit test for lognormality of the arsenic concentrations.

Null Hypothesis	Normal Scores (*G*)	Chi-Squared Scores (*C*)
Normal	`-2.62  (p=0.004)`	`30.2  (p=0.003)`
Lognormal	`-0.59  (p=0.28)`	`14.1  (p=0.29)`

Table 7.4. Results of Shapiro-Wilk group tests for the arsenic data.

Menu

To compute the Shapiro-Wilk group test statistics and create the companion probability plot, follow these steps.

1. In the Object Explorer, find and highlight the data frame **epa.92c.arsenic1.df**.
2. On the S-PLUS menu bar, make the following menu choices: **EnvironmentalStats>Hypothesis Tests>GOF Tests>Group> Shapiro-Wilk**. This will bring up the Shapiro-Wilk Group GOF Test dialog box.
3. In the Data Set box, make sure **epa.92c.arsenic1.df** is selected. In the Variable box select **Arsenic**, and in the Grouping Variable box select **Well**. In the Distribution box, make sure **Normal** is selected.
4. Click on the **Plotting** tab. Under Plot Type select the **Gi Q-Q** button, then click **OK** or **Apply**.

This creates the results for the test of a normal distribution. To create the results for the test of a lognormal distribution, in Step 3 select **Lognormal** in the Distribution box.

Command

To compute the Shapiro-Wilk group test statistics and create the companion probability plots, type these commands.

```
> attach(epa.92c.arsenic1.df)
> sw.list <- sw.group.gof(Arsenic, Well)
> sw.list
> plot(sw.list, plot.type="Gi Q-Q")
> sw.log.list <- sw.group.gof(Arsenic, Well,
    dist="lnorm")
> sw.log.list
> plot(sw.log.list, plot.type="Gi Q-Q")
> detach()
```

7.2.3   One Sample Goodness-of-Fit Tests for Other Distributions

Three other commonly used goodness-of-fit tests are the Kolmogorov-Smirnov goodness-of-fit test (Zar, 1999, p. 475), the chi-square goodness-of-fit test (Zar, 1999, p. 462), and the probability plot correlation coefficient (PPCC) goodness-of-fit test (Filliben, 1975; Vogel, 1986). The Kolmogorov-Smirnov and chi-square tests are built into S-PLUS, but the Kolmogorov-Smirnov test has been slightly modified in ENVIRONMENTALSTATS for S-PLUS to produce nicer output and plots. ENVIRONMENTALSTATS for S-PLUS also includes the PPCC test for the extreme value distribution. You can look at the help files for `ks.gof`, `chisq.gof`, and `ppcc.evd.gof` to get more detailed information and see examples.

In Chapter 3, Section 3.5.2, we created a Q-Q plot to determine whether a set of benzene concentrations appear to come from a Poisson distribution. Out of the 36 observations, 33 are reported as "<2", and these observations were set to half the detection limit (i.e., 1). The Kolmogorov-Smirnov test to formally test whether these data appear to come from a Poisson distribution yields a p-value that is essentially 0, so we have evidence that the assumption of a Poisson distribution is not valid.

Menu

To perform the Kolmogorov-Smirnov goodness-of-fit test for a Poisson distribution and plot a summary of the results, follow these steps.

1. In the Object Explorer, find and highlight the data frame **new.epa.92c.benzene1.df** that you created in Section 3.5.2.
2. On the S-PLUS menu bar, make the following menu choices: **EnvironmentalStats>Hypothesis Tests>GOF Tests>One Sample>Kolmogorov-Smirnov**. This will bring up the K-S GOF Test dialog box.
3. For Data to Use, select **Pre-Defined Data**. For Data Set make sure **new.epa.92c.benzene1.df** is selected, for Variable select **New.Benzene**. In the Distribution box select **Poisson**. **Check** the estimate parameters box.
4. Click **OK** or **Apply**.

Command

To perform the Kolmogorov-Smirnov goodness-of-fit test for a Poisson distribution and plot a summary of the results, type these commands.

```
> attach(new.epa.92c.benzene1.df)
> ks.list <- ks.gof(New.Benzene, distribution="pois")
> ks.list
> plot.gof.summary(ks.list)
> detach()
```

7.2.4   Two-Sample Goodness-of-Fit Test to Compare Samples

You can use the Kolmogorov-Smirnov test to test whether two sets of observations appear to come from the exact same distribution. In Chapter 3, Section

3.5.3, we created a Q-Q plot comparing the Reference area and Cleanup area TcCB concentrations based on the log-transformed data. The Kolmogorov-Smirnov test to formally test whether the two data sets appear to come from the same distribution yields a p-value of 0.013.

Menu

To perform the Kolmogorov-Smirnov goodness-of-fit test to compare the Reference and Cleanup areas and plot a summary of the results, follow these steps.

1. In the Object Explorer, find and highlight the data frame **new.epa.94b.tccb.df** that you created in Chapter 1, Section 1.11.3.
2. On the S-PLUS menu bar, make the following menu choices: **EnvironmentalStats>Hypothesis Tests>GOF Tests>Two Samples>Kolmogorov-Smirnov**. This will bring up the 2-Sample K-S GOF Test dialog box.
3. For Data Set select **new.epa.94b.tccb.df**. For Variable 1 select **log.TcCB**, for Variable 2 select **Area**, and **check** the box Variable 2 is a Grouping Variable.
4. Click **OK** or **Apply**.

Note that the result of the test is the same whether we use the original observations stored in the column TcCB or the log-transformed observations stored in the column log.TcCB, but the summary plots will be different.

Command

To perform the Kolmogorov-Smirnov goodness-of-fit test to compare the Reference and Cleanup areas and plot a summary of the results, type these commands.

```
> attach(epa.94b.tccb.df)
> log.TcCB.ref <- log(TcCB[Area=="Reference"])
> log.TcCB.clean <- log(TcCB[Area=="Cleanup"])
> ks.list <- ks.gof(log.TcCB.ref, log.TcCB.clean)
> ks.list
> plot.two.sample.gof.summary(ks.list)
> detach()
```

7.3   One-, Two-, and *k*-Sample Comparison Tests

Frequently in environmental studies, we are interested in comparing mean or median concentrations to a standard, or comparing mean or median concentrations between two or more areas (e.g., background versus potentially contaminated). Sometimes we are interested in a measure of variability as well. S-PLUS comes with several built-in menu items and functions for performing standard hypothesis tests for one-, two-, and *k*-sample comparisons (e.g., Student's t-test, analysis of variance, etc.). Table 7.5 lists additional menu items and functions available in ENVIRONMENTALSTATS for S-PLUS for comparing samples. All menu items fall under the menu choice **EnvironmentalStats>Hypothesis Tests>Compare Samples**. For example, the first entry in the table describes what you can do if you make the selection **EnvironmentalStats>Hypothesis Tests>Compare Samples>One Sample>Permutation Test**.

Menu Item	Functions	Description
One Sample> Permutation Test	`one.sample.permutation.test`	Fisher's one-sample permutation test for location.
One Sample> t-Test for Skewed Data	`chen.t.test`	Chen's modified one-sample t-test for skewed data.
One Sample> Sign Test	`sign.test`	Sign test.
One Sample> Chi-Square Test on Variance	`var.test`	One-sample chi-square test on variance.
Two Samples> Permutation Tests> Locations	`two.sample.permutation.test. location`	Two-sample or paired permutation test to compare locations.
Two Samples> Permutation Tests> Proportions	`two.sample.permutation.test. proportion`	Two-sample or paired permutation test to compare proportions.
Two Samples> Paired t-Test for Skewed Data	`chen.t.test`	Chen's modified paired t-test for skewed differences.
Two Samples> Linear Rank Test	`two.sample.linear.rank.test`	Two-sample linear rank test.
Two Samples> Paired Sign Test	`sign.test`	Paired sign test.
Two Samples> Quantile Test	`quantile.test`	Quantile test for a shift in the tail.
Two Samples> F Test for Equal Variances	`var.test`	F-test for equal variances between two populations.
k Samples> Homogeneous Variances	`var.group.test`	Levene's or Bartlett's test for equal variances among k populations.

Table 7.5. Menu items and functions in ENVIRONMENTALSTATS for S-PLUS for comparison tests.

All of the tests listed in Table 7.5, as well as examples of using the associated menu items and functions, are discussed in detail in Chapter 7 of Millard and Neerchal (2001). In this section we will give examples of performing Chen's modified t-test and comparing a linear rank test with the quantile test.

7.3.1   Chen's Modified One-Sample t-Test for Skewed Data

Student's t-test, Fisher's one-sample permutation test, and the Wilcoxon signed
rank test all assume that the underlying distribution is symmetric about its mean.
Chen (1995b) developed a modified t-statistic for performing a one-sided test of
hypothesis on the mean of a skewed distribution. For the case of a positively
skewed distribution, her test can be applied to the upper one-sided alternative
hypothesis (i.e., H_0: $\mu \le \mu_0$ versus H_a: $\mu > \mu_0$). For the case of a negatively
skewed distribution, her test can be applied to the lower one-sided alternative
hypothesis (i.e., H_0: $\mu \ge \mu_0$ versus H_a: $\mu < \mu_0$). Since environmental data are
usually positively skewed, her test would usually be applied to the case of testing the one-sided upper hypothesis.

The guidance document *Supplemental Guidance to RAGS: Calculating the
Concentration Term* (USEPA, 1992d) contains an example with 15 chromium
concentrations (mg/kg) which are assumed to come from a lognormal distribution. In ENVIRONMENTALSTATS for S-PLUS these data are stored in the vector
epa.92d.chromium.vec and the data frame epa.92d.chromium.df.

```
> epa.92d.chromium.vec
 [1]   10   13   20   36   41   59   67  110  110  136
[11]  140  160  200  230 1300
```

In this example we will use Chen's modified t-test to test the null hypothesis
that the average chromium concentration is less than 100 mg/kg versus the alternative that it is greater than 100 mg/kg. Here are the results:

```
Results of Hypothesis Test
--------------------------

Null Hypothesis:              mean = 100

Alternative Hypothesis:       True mean is greater than 100

Test Name:                    One-sample t-Test
                              Modified for
                              Positively-Skewed Distributions
                              (Chen, 1995)

Estimated Parameter(s):       mean = 175.4667
                              sd   = 318.544
                              skew =   3.572281

Data:                         epa.92d.chromium.vec

Test Statistic:               t = 1.562903

Test Statistic Parameter:     df = 14

P-values:                     z             = 0.05903767
                              t             = 0.07019591
                              Avg. of z and t = 0.06461679
```

```
Confidence Interval for:          mean

Confidence Interval Method:       Based on z

Confidence Interval Type:         Lower

Confidence Level:                 95%

Confidence Interval:              LCL = 96.89075
                                  UCL = Inf
```

The estimated mean, standard deviation, and skew are 175, 319, and 3.6, respectively. The p-value is 0.06, and the lower 95% confidence interval is [96.9, ∞). Depending on what you use for your Type I error rate, you may or may not want to reject the null hypothesis.

Menu

To perform Chen's modified t-test for the chromium data follow these steps.

1. In the Object Explorer, find and highlight **epa.92d.chromium.df**.
2. On the S-PLUS menu bar, make the following menu choices: **EnvironmentalStats>Hypothesis Tests>Compare Samples>One Sample>t-Test for Skewed Data**. This will bring up the One-Sample t-Test for Skewed Data dialog box.
3. For Data to Use select **Pre-Defined data**, for Data Set select **epa.92d.chromium.df**, and for Variable select **Cr**. In the Mean Under Null Hypothesis box, type **100**, in the Alternative Hypothesis box select **greater**, in the Confidence Level (%) box select **95**, and in the CI Method box select **z**, then click **OK** or **Apply**.

Command

To perform Chen's modified t-test for the chromium data type this command.

```
> chen.t.test(epa.92d.chromium.vec, mu=100)
```

7.3.2   Two-Sample Linear Rank Tests and the Quantile Test

The Wilcoxon rank sum test is an example of a two-sample linear rank test. A linear rank test can be written as follows:

$$L = \sum_{i=1}^{n_1} a\left(R_{1i}\right) \tag{7.1}$$

where n_1 denotes the number of observations in group 1, R_{1i} denotes the rank of the ith observation in group 1, and $a()$ is some function that is called the *score function*. A linear rank test is based on the sum of the scores for group 1. For the Wilcoxon rank sum test, the function $a()$ is simply the identity function. Other functions may work better at detecting a small shift in location, depending

on the shape of the underlying distributions. See the ENVIRONMENTALSTATS for
S-PLUS help file for `two.sample.linear.rank.test` for more information.

The Wilcoxon rank sum test and other linear rank tests for shifts in location
are all designed to detect a shift in the whole distribution of group 1 relative to
the distribution of group 2. Sometimes, we may be interested in detecting a difference between the two distributions where only a portion of the distribution of
group 1 is shifted relative to the distribution of group 2. The mathematical nota-
tion for this kind of shift is:

$$F_1(t) = (1-\varepsilon) \, F_2(t) + \varepsilon F_3(t) \, , \quad -\infty < t < \infty \tag{7.2}$$

where F_1 denotes the cumulative distribution function (cdf) of group 1, F_2 denotes the cdf of group 2, and ε denotes a fraction between 0 and 1. In the statistical literature, the distribution of group 1 is sometimes called a "contaminated"
distribution, because it is the same as the distribution of group 2, except it is partially contaminated with another distribution. If the distribution of group 1 is
partially shifted to the right of the distribution of group 2, F_3 denotes a cdf such
that

$$F_3(t) \le F_2(t) \, , \quad -\infty < t < \infty \tag{7.3}$$

with a strict inequality for at least one value of t. If the distribution of group 1 is
partially shifted to the left of the distribution of group 2, F_3 denotes a cdf such
that

$$F_3(t) \ge F_2(t) \, , \quad -\infty < t < \infty \tag{7.4}$$

with a strict inequality for at least one value of t.

The quantile test is a two-sample rank test to detect a shift in a proportion of
one population relative to another population (Johnson et al., 1987). Under the
null hypothesis, the two distributions are the same. If the alternative hypothesis
is that the distribution of group 1 is partially shifted to the right of the distribution of group 2, the test combines the observations, ranks them, and computes k,
which is the number of observations from group 1 out of the r largest observations. The test rejects the null hypothesis if k is too large.

In Chapter 1, Section 1.11.9 (pp. 23–25), we compared the Reference area
and Cleanup area TcCB concentrations using both the Wilcoxon rank sum test
and the quantile test. The Wilcoxon rank sum test yields a p-value of 0.88,
whereas the quantile test yields a p-value of 0.01. These results are not surprising, considering the histograms of the data shown in Figure 1.1 on page 10. The
steps for using the ENVIRONMENTALSTATS for S-PLUS menu to perform these
tests are given in Chapter 1 on page 24, and the commands to type for performing these tests are given on pages 24 and 25.

7.4   Testing for Serial Correlation

You can test for the presence of serial correlation in a time series or set of residuals from a linear fit using the ENVIRONMENTALSTATS for S-PLUS function `serial.correlation.test` or the menu selection **EnvironmentalStats>Hypothesis Tests>Serial Correlation**. You can test for the presence of lag-one serial correlation using either the rank von Neumann ratio test, the normal approximation based on the Yule-Walker estimate of lag-one correlation, or the normal approximation based on the MLE of lag-one correlation. Only the last method, however, allows for missing values in the time series. See Chapter 11 of Millard and Neerchal (2001) and the help file for `serial.correlation.test` for examples of testing for serial correlation.

7.5   Testing for Trend

Often in environmental studies we are interested in assessing the presence or absence of a long term trend. A parametric test for trend involves fitting a linear model that includes some measure of time as one of the predictor variables, and possibly allowing for serially correlated errors in the model. The ***Mann-Kendall test for trend*** (Mann, 1945) is a nonparametric test for trend that does not assume normally distributed errors. Hirsch et al. (1982) introduced a modification of this test they call the ***seasonal Kendall test***. This test allows for seasonality and possibly serially correlated observations as well. Parametric and nonparametric tests for trend are described in detail in Chapter 11 of Millard and Neerchal (2001).

Table 7.6 lists the menu items and functions available in ENVIRONMENTALSTATS for S-PLUS for nonparametric tests for trend. All menu items fall under the menu choice **EnvironmentalStats>Hypothesis Tests**. For example, the first entry in the table describes what you can do if you make the selection **EnvironmentalStats>Hypothesis Tests>Trend Tests>Kendall**.

Menu Item	Functions	Description
Trend Tests> Kendall	`kendall.trend.test`	Nonparametric test for monotonic trend based on Kendall's tau statistic.
Trend Tests> Seasonal Kendall	`seasonal.kendall.trend.test`	Nonparametric test for monotonic trend within each season based on Kendall's tau statistic. Allows for serial correlation as well.

Table 7.6. Menu items and functions in ENVIRONMENTALSTATS for S-PLUS for nonparametric tests for trend.

7.5.1   Testing for Trend in the Presence of Seasons

Figure 7.7 displays monthly estimated total phosphorus mass (mg) within a water column at station CB3.3e for the five-year time period October 1984 – September 1989 from a study on phosphorus concentration conducted in the Chesapeake Bay (Brunenmesiter, 1989; Neerchal and Brunenmesiter, 1993).   In
ENVIRONMENTALSTATS for S-PLUS these data are stored in the regular time series Total.P.rts and the data frame Total.P.df.

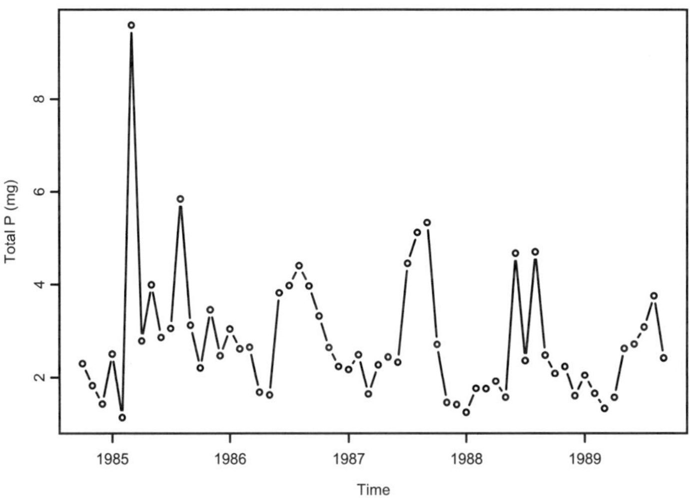

Figure 7.7. Monthly estimated total phosphorus mass (mg) within a water column at station CB3.3e in the Chesapeake Bay.

These data display seasonal variation, so we need to account for this while
testing for trend. Here are the results of the seasonal Kendall test for trend:

```
Results of Hypothesis Test
--------------------------


Null Hypothesis:              All 12 values of tau = 0

Alternative Hypothesis:       The seasonal taus are not all
                              equal
                              (Chi-Square Heterogeneity Test)

                              At least one seasonal tau != 0
                              and all non-zero tau's have the
                              same sign (z Trend Test)

Test Name:                    Seasonal Kendall Test for Trend
                              (with continuity correction)
```

```
Estimated Parameter(s):          tau       = -0.3333333
                                 slope     = -0.2312
                                 intercept = 346.5469

Estimation Method:               tau:        Weighted Average of
                                             Seasonal Estimates
                                 slope:      Hirsch et al.'s
                                             Modification of
                                             Thiel/Sen Estimator
                                 intercept:  Median of
                                             Seasonal Estimates

Data:                            y = CB3.3e

Sample Sizes:                    Oct   =  5
                                 Nov   =  5
                                 Dec   =  5
                                 Jan   =  5
                                 Feb   =  5
                                 Mar   =  5
                                 Apr   =  5
                                 May   =  5
                                 Jun   =  5
                                 Jul   =  5
                                 Aug   =  5
                                 Sep   =  5
                                 Total =  60

Test Statistics:                 Chi-Square (Het) =   4.48
                                 z (Trend)        = -2.757716

Test Statistic Parameter:        df = 11

P-values:                        Chi-Square (Het) = 0.953720141
                                 z (Trend)        = 0.005820666

Confidence Interval for:         slope

Confidence Interval Method:      Gilbert's Modification of
                                 Theil/Sen Method

Confidence Interval Type:        two-sided

Confidence Level:                95%

Confidence Interval:             LCL = -0.3616268
                                 UCL = -0.05569318
```

The estimated annual trend is –0.23 mg/year, i.e., a yearly decrease in total phosphorus. The p-value associated with the seasonal Kendall test for trend is $p = 0.006$, indicating this is statistically significant. The two-sided 95% confidence interval for the trend is [–0.36, –0.06]. The chi-square test for heterogeneity (i.e., is the trend different for different seasons?) yields a p-value of 0.95, so there is no evidence of different amounts of trend within different seasons.

Menu

To perform the seasonal Kendall test for trend, follow these steps.

1. Find **Total.P.df** in the Object Explorer.
2. On the S-PLUS menu bar, make the following menu choices: **EnvironmentalStats>Hypothesis Tests>Trend Tests>Seasonal Kendall**. This will bring up the Seasonal Kendall Trend Test dialog box.
3. For Data to Use select **Pre-Defined Data**. **Check** the Supply Season/Year box. The Data Set box should display **Total.P.df**. In the Variable box select **CB3.3e**, for Season select **Month**, and for Year select **Year**.
4. Click **OK** or **Apply**.

Command

To perform the seasonal Kendall test for trend, type these commands.

```
> attach(Total.P.df)
> seasonal.kendall.trend.test(CB3.3e, Month, Year)
> detach()
```

7.6 Summary

- S-PLUS contains several menu items and functions for performing classical statistical hypothesis tests, such as t-tests, analysis of variance, linear regression, nonparametric tests, quality control procedures, and time series analysis (see the S-PLUS documentation and help files).
- ENVIRONMENTALSTATS for S-PLUS contains menu items and functions for some statistical tests that are not included in S-PLUS but that are often used in environmental statistics
- Table 7.1 lists menu items and functions available in ENVIRONMENTALSTATS for S-PLUS for performing goodness-of-fit tests.
- Table 7.5 lists menu items and functions available in ENVIRONMENTALSTATS for S-PLUS for performing one-, two- and k-sample comparison tests.
- You can test for the presence of serial correlation in a time series or set of residuals from a linear fit using the ENVIRONMENTALSTATS for S-PLUS function `serial.correlation.test` or the menu selection **EnvironmentalStats>Hypothesis Tests>Serial Correlation**.
- Table 7.6 lists the menu items and functions available in ENVIRONMENTALSTATS for S-PLUS for performing nonparametric tests for trend.

8

Censored Data

8.1   Introduction

Often in environmental data analysis values are reported simply as being "below detection limit" along with the stated detection limit (e.g., USEPA, 1992c, p. 25; Porter et al., 1988). A sample of data contains **censored observations** if some of the observations are reported only as being below or above some censoring level. Although this results in some loss of information, we can still use data that contain nondetects for graphical and statistical analyses. Statistical methods for dealing with censored data have a long history in the field of survival analysis and life testing (see the help file *References: Censored Data – General References*). In this chapter, we will discuss how to create graphs, estimate distribution parameters and quantiles, construct prediction and tolerance intervals, perform goodness-of-fit tests, and compare distributions using censored data. See Chapter 10 of Millard and Neerchal (2001) for a more in-depth discussion of analyzing censored data.

8.2   Classification of Censored Data

There are four major ways to classify censored data: truncated versus censored, left versus right versus double, single versus multiple (progressive), and censored Type I versus censored Type II (Cohen, 1991, pp. 3–5). Most environmental data sets with nondetect values are either Type I left singly censored or Type I left multiply censored.

A sample of N observations is **left singly censored** (also called singly censored on the left) if c observations are known only to fall below a known censoring level T, while the remaining n ($n = N-c$) uncensored observations falling above T are fully measured and reported.

A sample is **singly censored** (e.g., singly left censored) if there is only one censoring level T. A sample is **multiply censored** or progressively censored (e.g., multiply left censored) if there are several censoring levels T_1, T_2, ..., T_p, where $T_1 < T_2 < ... < T_p$.

A censored sample has been subjected to **Type I censoring** if the censoring level(s) is(are) known in advance, so that given a fixed sample size N, the number of censored observations c (and hence the number of uncensored observations n) is a random outcome. Type I censored samples are sometimes called time-censored samples (Nelson, 1982, p. 248).

8.3   Menu Items and Functions for Censored Data

Table 8.1 lists the menu items and functions available in ENVIRONMENTALSTATS for S-PLUS for analyzing censored data. All menu items fall under the menu choice **EnvironmentalStats>Censored Data**. For example, the first entry in the table describes what you can do if you make the selection **EnvironmentalStats>Censored Data>EDA>CDF Plot>Empirical CDF**.

Menu Item	Functions	Description
EDA> CDF Plot> Empirical CDF	`ecdfplot.censored`	Empirical CDF based on censored data.
EDA> CDF Plot> Compare Two CDFs	`cdf.compare.censored`	Compare an empirical CDF to a hypothesized CDF, or compare two CDFs, based on censored data.
EDA> Q-Q Plot	`qqplot.censored`	Q-Q plot based on censored data.
EDA> Box-Cox Transformations	`boxcox.singly.censored` `boxcox.multiply.censored`	Determine an optimal Box-Cox transformation based on censored data.
Estimation> Parameters	`e`*abb*`.singly.censored` `e`*abb*`.multiply.censored`	Estimate the parameters of the distribution with the abbreviation *abb* based on censored data, and optionally construct a confidence interval for the parameters.
Hypothesis Tests> GOF Tests> Shapiro-Wilk	`sw.singly.censored.gof`	Shapiro-Wilk goodness-of-fit test for normality based on singly censored data.
Hypothesis Tests> GOF Tests> Shapiro-Francia	`sf.singly.censored.gof` `sf.multiply.censored.gof`	Shapiro-Francia goodness-of-fit test for normality based on censored data.
Hypothesis Tests> GOF Tests> PPCC	`ppcc.norm.singly.censored.gof` `ppcc.norm.multiply.censored.gof`	PPCC goodness-of-fit test for normality based on censored data.
Hypothesis Tests> Compare Two Samples	`two.sample.linear.rank.test.` `censored`	Two-sample linear rank test based on censored data.

Table 8.1. Menu items and functions in ENVIRONMENTALSTATS for S-PLUS for analyzing censored data.

8.4   Graphical Assessment of Censored Data

In Chapter 3 we illustrated several ways of creating graphs for a single variable, including histograms, quantile (empirical cdf) plots, and probability (Q-Q) plots. When you have censored data, creating a histogram is not necessarily straightforward (especially with multiply censored data), but you can create quantile plots and probability plots, as well as determine "optimal" Box-Cox transformations (see the first four rows of Table 8.1).

8.4.1   Quantile (Empirical CDF) Plots for Censored Data

In Chapter 3 we explained that a quantile plot (also called an empirical cumulative distribution function plot or empirical cdf plot) plots the ordered data (the empirical quantiles) on the x-axis versus the estimated cumulative probabilities (or plotting positions) on the y-axis. Various formulas for the plotting positions are given in Chapter 3 of Millard and Neerchal (2001) and the help file for `ecdfplot`. When you have censored data, the formulas for the plotting positions must be modified. For right-censored data, various formulas for the plotting positions are given by Kaplan and Meier (1958), Nelson (1972), and Michael and Schucany (1986). For left-censored data, formulas for the plotting positions are given by Michael and Schucany (1986) and Hirsch and Stedinger (1987).

When you have Type I left-censored data with only one censoring level, and all of the uncensored observations are larger than the censoring level, the computation of the plotting positions is straightforward because it is easy to order the uncensored observations. When you have one or more uncensored observations with values less than one or more of the censoring levels, then the computation of the plotting positions becomes a bit trickier. The help file for `ppoints.censored` in ENVIRONMENTALSTATS for S-PLUS gives a detailed explanation of the formulas for the plotting positions for censored data.

Table 8.2 displays 56 silver concentrations (μg/L) from an interlab comparison that include 34 values below one of 12 detection limits (Helsel and Cohn, 1988). In ENVIRONMENTALSTATS for S-PLUS these data are stored in the data frame `helsel.cohn.88.silver.df`.

Silver Concentrations (μg/L)								
<0.1	<0.1	0.1	0.1	<0.2	<0.2	<0.2	<0.2	0.2
<0.3	<0.5	0.7	0.8	<1	<1	<1	<1	<1
<1	<1	<1	<1	<1	1	1	1	1.2
1.4	1.5	<2	2	2	2	2	<2.5	2.7
3.2	4.4	<5	<5	<5	<5	5	<6	<10
<10	<10	<10	<10	10	<20	<20	<20	<25
90	560							

Table 8.2. Silver concentrations from an interlab comparison (Helsel and Cohn, 1988)

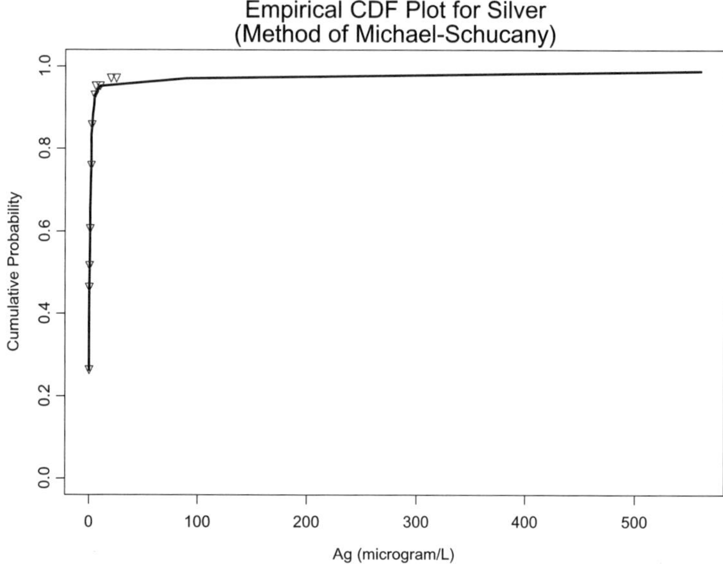

Figure 8.1. Empirical cdf plot of the silver data.

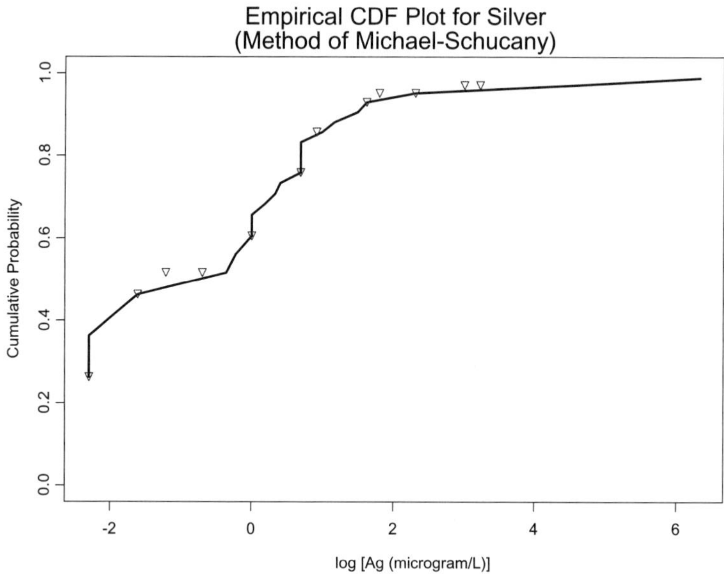

Figure 8.2. Empirical cdf plot of the log-transformed silver data.

Figure 8.1 displays the empirical cdf plot for the silver data. This plot indicates the data are extremely skewed to the right. This is not surprising since looking at the data in the table we see that all of the observations are less than 25 µg/L except for two that are 90 and 560 µg/L. Figure 8.2 displays the quantile plot based on the log-transformed observations. In both of these plots, the upside-down triangles indicate the censoring levels of observations that have been censored.

Menu

To create the quantile plot for the silver data follow these steps.

1. Find the data frame **helsel.cohn.88.silver.df** in the Object Explorer.
2. On the S-PLUS menu bar, make the following menu choices: **EnvironmentalStats>Censored Data>EDA>CDF Plot>Empirical CDF**. This will bring up the Plot Empirical CDF for Censored Data dialog box.
3. The Data Set box should display **helsel.cohn.88.silver.df**. For Variable select **Ag** and for Censor Ind. select **Censored**.
4. Click on the **Plot Options** tab, and **check** the Include Censored Values box, then click **OK** or **Apply**.

To create the quantile plot for the log-transformed silver data, repeat the above steps but in Step 3 for Variable select **log.Ag**.

Command

To create the quantile plots for the original and log-transformed silver data type these commands.

```
> attach(helsel.cohn.88.silver.df)
> ecdfplot.censored(Ag, Censored,
    xlab="Ag (microgram/L)", main="Empirical CDF Plot for
    Silver\n(Method of Michael-Schucany)", include.cen=T)
> ecdfplot.censored(log.Ag, Censored,
    xlab="log [Ag (microgram/L)]", main="Empirical CDF
    Plot for Silver\n(Method of Michael-Schucany)",
    include.cen=T)
> detach()
```

8.4.2   Comparing an Empirical and Hypothesized CDF

Figure 8.3 compares the empirical cdf of the silver data with a lognormal cdf, where the parameters for the lognormal distribution are estimated from the data (see the section *Estimating Distribution Parameters* later in this chapter). Figure 8.4 compares the log-transformed silver data with a normal distribution.

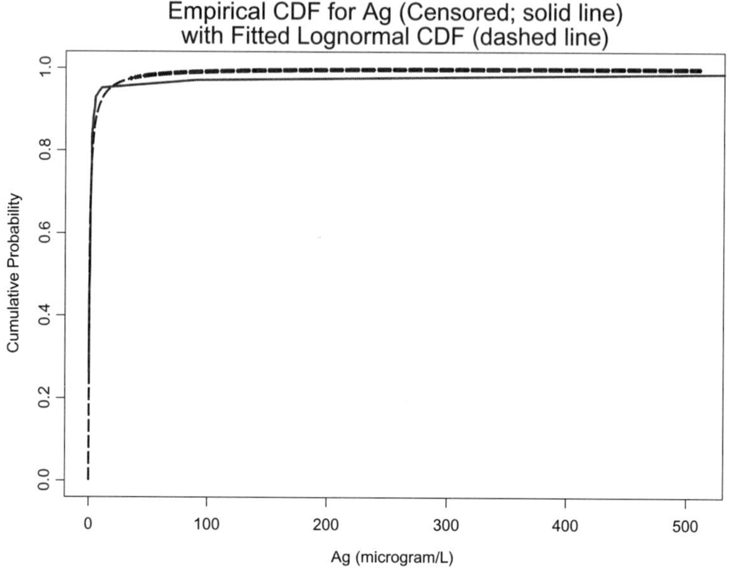

Figure 8.3. Empirical cdf of the silver data with a fitted lognormal distribution.

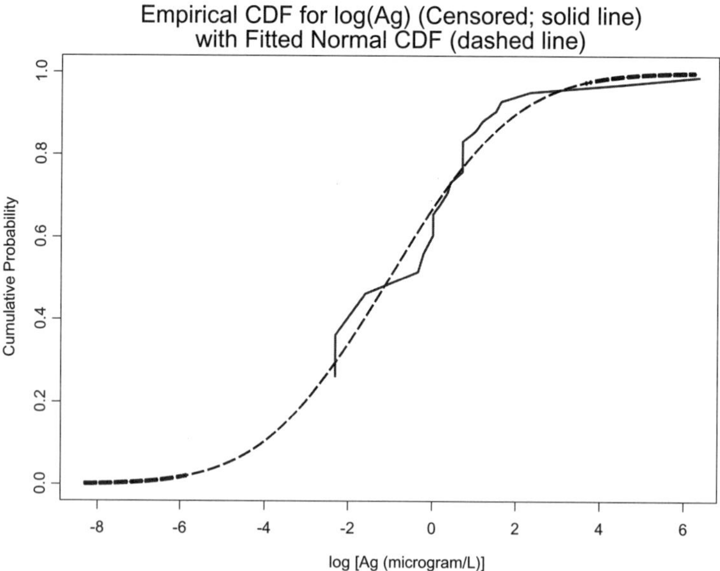

Figure 8.4. Empirical cdf of the log-transformed silver data with a fitted normal distribution.

The plots appear to show that the lognormal distribution provides an adequate fit to these data but the Shapiro-Francia goodness-of-fit test for lognormality yields a p-value of 0.03 (see the section *Goodness-of-Fit Tests* later in this chapter).

Menu

To create the plot in Figure 8.3 follow these steps.

1. Find **helsel.cohn.88.silver.df** in the Object Explorer.
2. On the S-PLUS menu bar, make the following menu choices: **EnvironmentalStats>Censored Data>EDA>CDF Plot>Compare Two CDFs**. This will bring up the Compare Two CDFs for Censored Data dialog box.
3. For Compare Data to select **Distribution**. The Data Set box should display **helsel.cohn.88.silver.df**. For Variable 1 select **Ag** and for Censor Ind. 1 select **Censored**. Under the Distribution Information Group, make sure that the **Estimate Parameters** box is checked. For Distribution select **Lognormal**.
4. Click **OK** or **Apply**.

To create the plot in Figure 8.4 follow the same steps as above, but in Step 2 for Variable 1 select **log.Ag** and for Distribution select **Normal**.

Command

To create the plots in Figure 8.3 and Figure 8.4 type these commands.

```
> attach(helsel.cohn.88.silver.df)
> cdf.compare.censored(Ag, Censored,
    distribution="lnorm", xlab="Ag (microgram/L)")
> cdf.compare.censored(log(Ag), Censored,
    distribution="norm", xlab="log [Ag (microgram/L)]")
> detach()
```

8.4.3   Comparing Two Empirical CDFs

Table 8.3 displays copper concentrations (μg/L) in shallow groundwater samples from two different geological zones in the San Joaquin Valley, California (Millard and Deverel, 1988). The alluvial fan data include four different detection limits and the basin trough data include five different detection limits. In ENVIRONMENTALSTATS for S-PLUS these data are stored in the data frame `millard.deverel.88.df`. Figure 8.5 compares the empirical cdf of copper concentrations from the alluvial fan zone with those from the basin trough zone. This plot shows that the two distributions are fairly similar in shape and location.

Zone	Copper (µg/L)										
Alluvial Fan	<1	<1	<1	<1	1	1	1	1	1	2	2
	2	2	2	2	2	2	2	2	2	2	2
	2	2	2	2	2	2	2	2	3	3	3
	3	3	3	4	4	4	<5	<5	<5	<5	<5
	<5	<5	<5	5	5	5	7	7	7	8	9
	<10	<10	<10	10	11	12	16	<20	<20	20	NA
	NA	NA									
Basin Trough	<1	<1	1	1	1	1	1	1	1	<2	<2
	2	2	2	2	3	3	3	3	3	3	3
	3	4	4	4	4	4	<5	<5	<5	<5	<5
	5	6	6	8	9	9	<10	<10	<10	<10	12
	14	<15	15	17	23	NA					

Table 8.3. Copper concentrations in shallow groundwater in two geological zones.

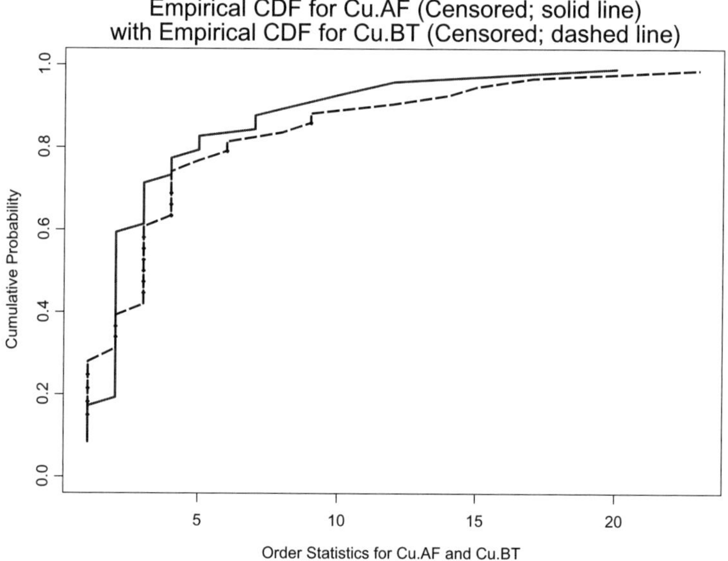

Figure 8.5. Empirical CDFs of copper concentrations in the alluvial fan and basin trough zones.

Menu

To create the plot in Figure 8.5 follow these steps.

1. Find **millard.deverel.88.df** in the Object Explorer.

2. On the S-PLUS menu bar, make the following menu choices: **EnvironmentalStats>Censored Data>EDA>CDF Plot>Compare Two CDFs**. This will bring up the Compare Two CDFs for Censored Data dialog box.

3. For Compare Data select **Other Data**. The Data Set box should display **millard.deveral.88.silver.df**. For Variable 1 choose **Cu**, for Censor Ind. 1 choose **Cu.censored**, for Variable 2 choose **Zone**, **check** the Variable 2 is a Grouping Variable box.

4. Click **OK** or **Apply**.

Command

To create the plot in Figure 8.5 type these commands.

```
> attach(millard.deverel.88.df)
> Cu.AF <- Cu[Zone=="Alluvial.Fan"]
> Cu.AF.cen <- Cu.censored[Zone=="Alluvial.Fan"]
> Cu.BT <- Cu[Zone=="Basin.Trough"]
> Cu.BT.cen <- Cu.censored[Zone=="Basin.Trough"]
> cdf.compare.censored(Cu.AF, Cu.AF.cen, "left", Cu.BT,
     Cu.BT.cen, "left")
> detach()
```

8.4.4   Q-Q Plots for Censored Data

In Chapter 3 we explained that a probability or quantile-quantile (Q-Q) plot plots the ordered data (the empirical quantiles) on the y-axis versus the corresponding quantiles from the assumed theoretical probability distribution on the x-axis, where the quantiles from the assumed distribution are computed based on the plotting positions. As for empirical cdf plots, when you have censored data, the formulas for the plotting positions must be modified.

Table 8.4 presents artificial TcCB concentrations based on the Reference area data. For this data set, the concentrations of TcCB less than 0.5 ppb have been recoded as "<0.5," so there are with 19 censored observations, 28 uncensored observations, and a total sample size of 47. In ENVIRONMENTALSTATS for S-PLUS these data are stored in the data frame `Modified.TcCB.df`.

TcCB Concentrations (ppb)							
<0.5	<0.5	<0.5	<0.5	<0.5	<0.5	<0.5	<0.5
<0.5	<0.5	<0.5	<0.5	<0.5	<0.5	<0.5	<0.5
<0.5	<0.5	<0.5	0.5	0.5	0.51	0.52	0.54
0.56	0.56	0.57	0.57	0.6	0.62	0.63	0.67
0.69	0.72	0.74	0.76	0.79	0.81	0.82	0.84
0.89	1.11	1.13	1.14	1.14	1.2	1.33	

Table 8.4. Modified Reference area TcCB concentrations.

Figure 8.6 shows the normal Q-Q plot for the log-transformed modified TcCB data. This plot indicates that the lognormal distribution appears to provide an adequate fit to these data. This is not surprising since we already saw in Figure 1.6 on page 17 that a lognormal distribution provides a good fit to the original data.

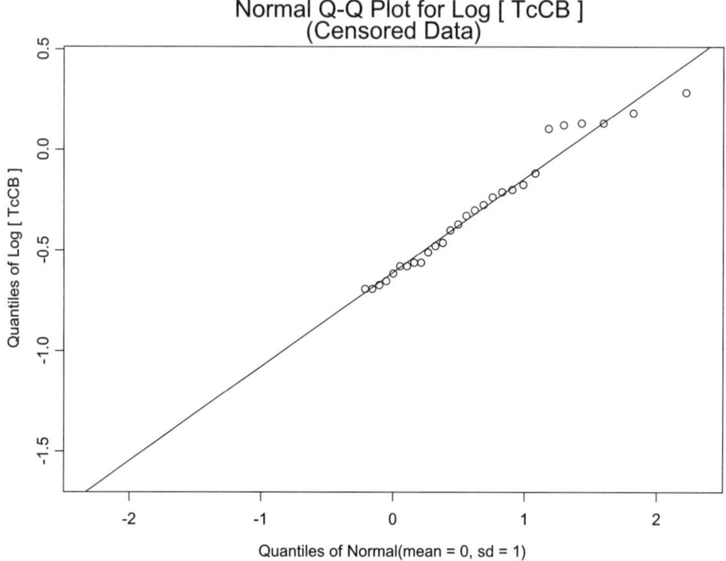

Figure 8.6. Normal Q-Q plot for the log-transformed singly left-censored TcCB data of Table 8.4.

Menu

To create the Q-Q plot for the modified TcCB data using the ENVIRONMENTALSTATS for S-PLUS pull-down menu, follow these steps.

1. Find **Modified.TcCB.df** in the Object Explorer.
2. On the S-PLUS menu bar, make the following menu choices: **EnvironmentalStats>Censored Data>EDA>Q-Q Plot**. This will bring up the Q-Q Plot for Censored Data dialog box.
3. The Data Set box should display **Modified.TcCB.df**. For Variable 1 choose **TcCB**, for Censor Ind. 1 select **Censored**, and for Distribution select **Lognormal**.
4. Click on the **Plotting** tab. **Click** on the Add a Line box to select this option. Click **OK** or **Apply**.

Command

To create the Q-Q plot for the modified TcCB data type these commands.

```
> attach(Modified.TcCB.df)
```

```
> qqplot.censored(TcCB, Censored, distribution="lnorm",
    add.line=T)
> detach()
```

8.4.5   Box-Cox Transformations for Censored Data

In Chapter 3 we discussed using Box-Cox transformations as a way to satisfy normality assumptions for standard statistical tests, and also sometimes to satisfy the linear assumption and/or the constant variance in the errors assumption for a standard linear regression model. We also discussed three possible criteria to use to decide on the power of the transformation: the probability plot correlation coefficient (PPCC), the Shapiro-Wilk goodness-of-fit test statistic, and the log-likelihood function. This idea can be extended to the case of singly and multiply censored data (e.g., Shumway et al., 1989). See the help files for `boxcox.singly.censored` and `boxcox.multiply.censored` for details.

Figure 8.7 displays a plot of the probability plot correlation coefficient versus various values of the transform power λ for the singly censored TcCB data shown in Table 8.4. For these data, the PPCC reaches its maximum between about $\lambda = 0$ (log transformation) and $\lambda = 0.5$ (square-root transformation). We saw a similar pattern for the original Reference area TcCB data in Figure 3.7 on page 72, although in that figure the maximum appeared at about $\lambda = 0$.

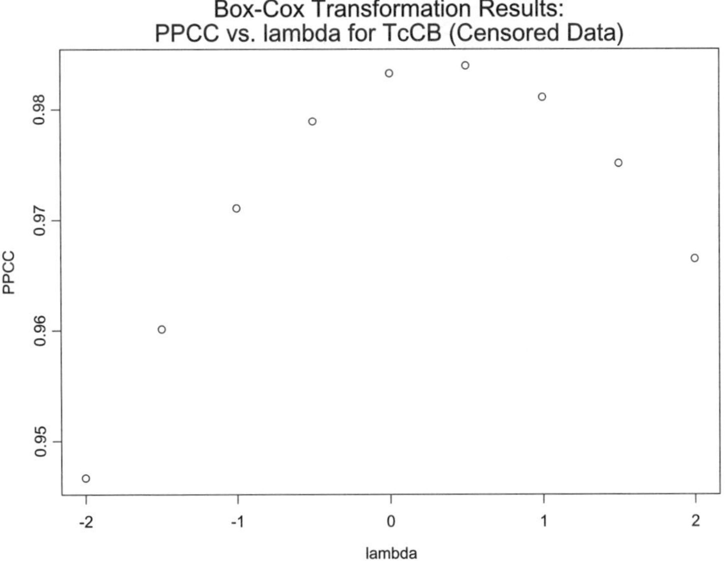

Figure 8.7. Probability plot correlation coefficient versus Box-Cox transform power (λ) for the singly censored TcCB data of Table 8.4.

Menu

To create the plot of the PPCC versus the transformation power for the modified TcCB data, follow these steps.

1. Find **Modified.TcCB.df** in the Object Explorer.
2. On the S-PLUS menu bar, make the following menu choices: **EnvironmentalStats>Censored Data>EDA>Box-Cox Transformations**. This will bring up the Box-Cox Transformations for Censored Data dialog box.
3. For Data to Use make sure the **Pre-Defined Data** button is selected. The Data Set box should display **Modified.TcCB.df**. For Variable choose **TcCB**, and for Censor Ind. select **Censored**.
4. Click **OK** or **Apply**.

Command

To create the plot of the PPCC versus the transformation power for the modified TcCB data, type these commands.

```
> attach(Modified.TcCB.df)
> boxcox.list <- boxcox.singly.censored(TcCB, Censored)
> plot(boxcox.list, plot.type="Objective vs. lambda")
> detach()
```

8.5   Estimating Distribution Parameters

In Chapter 5 we illustrated how to use ENVIRONMENTALSTATS for S-PLUS to estimate distribution parameters and quantiles, and also create confidence intervals for these quantities. Methods for estimating parameters include maximum likelihood, method of moments, and minimum variance unbiased. It is fairly straightforward to extend maximum likelihood estimation to the case of censored data (e.g., Cohen, 1991; Schneider, 1986). More recently, researchers in the environmental field have proposed alternative methods of computing estimates and confidence intervals in addition to the classical ones such as maximum likelihood estimation. See Chapter 10 of Millard and Neerchal (2001) for details.

8.5.1   Estimating Mean TcCB Concentrations

In Section 5.2.2 on page 97 we estimated the mean TcCB concentration in the Reference area as 0.6 ppb and the CV as 0.49, assuming the data come from a lognormal distribution. We also computed a two-sided 95% confidence intervals for the mean as [0.52, 0.7]. There are 47 uncensored observations in this data set. The modified TcCB data shown in Table 8.4 include 19 censored observations. Using these data, the estimated mean and CV are 0.61 and 0.46

based on the maximum likelihood method, and the two-sided 95% confidence interval for the mean is [0.51, 0.73] based on Cox's method.

Menu

To estimate the mean and coefficient of variation, and compute the two-sided 95% confidence interval for the mean, follow these steps.

1. Find **Modified.TcCB.df** in the Object Explorer.
2. On the S-PLUS menu bar, make the following menu choices: **EnvironmentalStats>Censored Data>Estimation>Parameters**. This will bring up the Estimate Distribution Parameters for Censored Data dialog box.
3. For Data to Use select **Pre-Defined Data**. The Data Set box should display **Modified.TcCB.df**. For Variable select **TcCB**, for Censor Ind. Select **Censored**, for Distribution select **Lognormal (Alternative)**, and for Estimation Method select **mle**. Under the Confidence Interval section **check** the Confidence Interval box, for CI Type select **two-sided**, for CI Method select **cox**, for Confidence Level (%) select **95**, for Pivot Stat. select **t**.
4. Click **OK** or **Apply**.

Command

To estimate the mean and coefficient of variation, and compute the two-sided 95% confidence interval for the mean, type these commands.

```
> attach(Modified.TcCB.df)
> elnorm.alt.singly.censored(TcCB, Censored,
    method="mle", ci=T, ci.method="cox",
    pivot.statistic="t")
> detach()
```

8.6   Estimating Distribution Quantiles

In Chapter 5 we illustrated how to estimate and construct confidence intervals for population quantiles or percentiles, both parametrically and nonparametrically. In this section we will discuss how to do this with censored data.

8.6.1   Parametric Estimates of Quantiles

For a normal distribution, the estimated quantile and confidence interval are functions of the estimated mean and standard deviation. For a lognormal distribution, they are functions of the estimated mean and standard deviation based on the log-transformed observations. In the presence of censored observations, we can construct parametric estimates and confidence intervals for population percentiles using the same formulas as for complete data, but to estimate the centiles using the same formulas as for complete data, but to estimate the mean and standard deviation we have to use the formulas that are appropriate for censored data. It is not clear how well the methods for estimating and constructing confidence intervals for population percentiles behave in the presence of censored data. This is an area that requires further research.

The estimated 90th percentile of the Reference area TcCB concentrations is 0.98 ppb, assuming these data come from a lognormal distribution. Also, the two-sided 95% confidence interval for the 90th percentile is [0.84, 1.22]. If we instead use the modified TcCB concentrations shown in Table 8.4, the estimated 90th percentile is 0.97 and the two-sided 95% confidence interval for the 90th percentile is [0.84, 1.19].

Menu

To estimate the 90th percentile of the Reference area TcCB concentrations and construct a two-sided 95% confidence interval for the 90th percentile using the modified TcCB data, follow these steps.

1. Find **Modified.TcCB.df** in the Object Explorer.
2. On the S-PLUS menu bar, make the following menu choices: **EnvironmentalStats>Censored Data>Estimation>Parameters**. This will bring up the Estimate Distribution Parameters for Censored Data dialog box.
3. For Data to Use select **Pre-Defined Data**. The Data Set box should display **Modified.TcCB.df**. For Variable select **TcCB**, for Censor Ind. Select **Censored**, for Distribution select **Lognormal**, and for Estimation Method select **mle**. Under the Confidence Interval section **uncheck** the Confidence Interval box. **Uncheck** the Print Results box. In the Save As box type **estimation.list**.
4. Click **OK**.
5. On the S-PLUS menu bar, make the following menu choices: **EnvironmentalStats>Estimation>Quantiles**. This will bring up the Estimate Distribution Quantiles dialog box.
6. For Data to Use select **Expression**. In the Expression box type **estimation.list**. In the Quantile(s) box select **0.90**. For Distribution select **Lognormal**. Under the Confidence Interval group **check** the Confidence Interval box. For CI Type select **two-sided**. For CI Method select **exact**. For Confidence Level (%) select **95**.
7. Click **OK** or **Apply**.

Command

To estimate the 90th percentile of the Reference area TcCB concentrations and construct a two-sided 95% confidence interval for the 90th percentile using the modified TcCB data, type these commands.

```
> attach(Modified.TcCB.df)
```

```
> estimation.list <- elnorm.singly.censored(TcCB,
    Censored, method="mle", ci=F)
> eqlnorm(estimation.list, p=0.9, ci=T)
> detach()
```

8.6.2   Nonparametric Estimates of Quantiles

We saw in Chapter 5 that nonparametric estimates and confidence intervals for population percentiles are simply functions of the ordered observations. Thus, for left censored data, you can still estimate quantiles and create one-sided upper confidence intervals as long as there are enough uncensored observations that can be ordered in a logical way, just as we did in Section 5.3.3 using the copper data shown in Table 5.5.

8.7   Prediction and Tolerance Intervals

In Chapter 6 we illustrated how to construct parametric and nonparametric prediction and tolerance intervals. In this section we will discuss how to do this with censored data.

8.7.1   Parametric Prediction and Tolerance Intervals

For a normal distribution, prediction and tolerance intervals are functions of the estimated means and standard deviations. For a lognormal distribution, they are functions of the estimated mean and standard deviation based on the log-transformed observations. In the presence of censored observations, we can construct prediction and tolerance intervals using the same formulas as for complete data, but to estimate the mean and standard deviation we have to use the formulas that are appropriate for censored data. It is not clear how well the methods for constructing prediction and tolerance intervals behave in the presence of censored data. This is an area that requires further research

In Section 6.4.2 on page 127 we compared a one-sided upper tolerance limit with a one-sided upper prediction limit based on using the Reference area TcCB data and assuming a lognormal distribution. The tolerance limit was a 95% β-content tolerance limit with associated confidence level of 95%. The prediction limit was for the next $k = 77$ observations with associated confidence level of 95%. The tolerance limit was 1.42 and the prediction limit was 2.68. For the singly censored data set shown in Table 8.4, these limits become 1.38 and 2.51. In this case there is very little difference between the limits based on the complete data versus those based on the censored data.

Menu

To compute the one-sided upper 95% β-content tolerance limit with associated confidence level 95% based on the modified TcCB data, follow these steps.

1. Find **Modified.TcCB.df** in the Object Explorer.
2. On the S-PLUS menu bar, make the following menu choices: **EnvironmentalStats>Censored Data>Estimation>Parameters**. This will bring up the Estimate Distribution Parameters for Censored Data dialog box.
3. For Data to Use select **Pre-Defined Data**. The Data Set box should display **Modified.TcCB.df**. For Variable select **TcCB**, for Censor Ind. Select **Censored**, for Distribution select **Lognormal**, and for Estimation Method select **mle**.
4. Under the Confidence Interval section **uncheck** the Confidence Interval box, in the Save As box type **estimation.list**, then click **OK**.
5. On the S-PLUS menu bar, make the following menu choices: **EnvironmentalStats>Estimation>Tolerance Intervals**. This will bring up the Tolerance Interval dialog box.
6. For Data to Use select **Expression**. In the Expression box type **estimation.list**.
7. Click on the **Interval** tab. Under the Distribution group, for Type select **Parametric**, and for Distribution select **Lognormal**. Under the Tolerance Interval group, for Coverage Type select **Content**, for TI Type select **upper**, for Coverage (%) select **95**, and for Confidence Level (%) type **95**.
8. Click **OK** or **Apply**.

To compute the one-sided upper prediction limit, follow these steps.

1. On the S-PLUS menu bar, make the following menu choices: **EnvironmentalStats>Estimation>Prediction Intervals**. This will bring up the Prediction Interval dialog box.
2. For Data to Use select **Expression**. In the Expression box type **estimation.list**.
3. Click on the **Interval** tab. Under the Distribution/Sample Size group, for Type select **Parametric** and for Distribution select **Lognormal**. Under the Prediction Interval group, **uncheck** the Simultaneous box, for # Future Obs type **77**, for PI Type select **upper**, for PI Method select **exact**, and for Confidence Level (%) select **95**.
4. Click **OK** or **Apply**.

Command

To compute the one-sided upper 95% β-content tolerance limit with associated confidence level 95% based on the modified TcCB data, type these commands.

```
> attach(Modified.TcCB.df)
```

```
> estimation.list <- elnorm.singly.censored(TcCB,
    Censored, method="mle", ci=F)
> tol.int.lnorm(estimation.list, coverage=0.95,
    cov.type="content", ti.type="upper", conf.level=0.95)
```

To compute the one-sided upper prediction limit, type these commands.

```
> pred.int.lnorm(estimation.list, k=77, method="exact",
    pi.type="upper", conf.level=0.95)
> detach()
```

8.7.2   Nonparametric Prediction and Tolerance Intervals

We saw in Chapter 6 that nonparametric prediction and tolerance intervals are simply functions of the ordered observations. Thus, for left censored data, you can still create nonparametric prediction and tolerance intervals as long as there are enough uncensored observations that can be ordered in a logical way (see the examples in Section 6.2.3 and Section 6.4.3).

8.8   Hypothesis Tests

In this section we will discuss how to perform hypothesis tests in the presence of censored data.

8.8.1   Goodness-of-Fit Tests

Royston (1993) extended both the Shapiro-Francia and Shapiro-Wilk goodness-of-fit tests to the case of singly censored data. He also provides a method of computing p-values for these statistics based on tables given in Verrill and Johnson (1988). For details, see the help files for `sw.singly.censored.gof` and `sf.singly.censored.gof`. Although Verrill and Johnson (1988) produced their tables based on Type II censoring, Royston's (1993) approximation to the p-value of these tests should be fairly accurate for Type I censored data as well.

The Shapiro-Francia and PPCC tests are also easily extendible to the case of multiply censored data, but it is not known how well Royston's (1993) method of computing p-values works in this case. See the ENVIRONMENTALSTATS for S-PLUS help file for `sf.multiply.censored.gof` for details.

We noted in Chapter 7 that goodness-of-fit tests are of limited value for small sample sizes because there is usually not enough information to distinguish between different kinds of distributions. This also holds true even for moderate sample sizes if you have censored data, as the next example shows.

In Section 7.2.1 we showed that goodness-of-fit tests on the Reference area TcCB data indicated that the normal distribution was not appropriate, but that a lognormal distribution appeared to adequately model the data. In this example we will perform the same tests but use the modified TcCB data of Table 8.4. The Shapiro-Wilk test for normality yields a p-value of 0.34 and the test for lognormality yields a p-value of 0.28. Figure 8.8 and Figure 8.9 show companion plots for the tests for normality and lognormality, respectively. Compare these figures to Figure 7.1 and Figure 7.2. We see that unlike the case with complete data, here censoring 40% of the observations leaves us unable to distinguish between a normal and lognormal distribution.

Menu

To perform the Shapiro-Wilk test for normality using the ENVIRONMENTALSTATS for S-PLUS pull-down menu, follow these steps.

1. Find **Modified.TcCB.df** in the Object Explorer.
2. On the S-PLUS menu bar, make the following menu choices: **EnvironmentalStats>Censored Data>Hypothesis Tests>GOF Tests>Shapiro-Wilk**. This will bring up the Shapiro-Wilk GOF Test for Censored Data dialog box.
3. For Data Set make sure **Modified.TcCB.df** is selected, for Variable select **TcCB**, and in Censor Ind. select **Censored**. In the Distribution box select **Normal**.
4. Click on the **Plotting** tab. Under Plotting Information, in the Significant Digits box type **3**, then click **OK** or **Apply**.

To perform the Shapiro-Wilk test for lognormality, repeat the above steps, but in Step 3 select **Lognormal** in the Distribution box.

Command

To perform the Shapiro-Wilk tests type these commands.

```
> attach(Modified.TcCB.df)
> sw.list.norm <- sw.singly.censored.gof(TcCB, Censored)
> sw.list.norm
> sw.list.lnorm <- sw.singly.censored.gof(TcCB, Censored,
    dist="lnorm")
> sw.list.lnorm
```

To plot the results of these tests as shown in Figure 8.8 and Figure 8.9 , type these commands.

```
> plot.gof.censored.summary(sw.list.norm, digits=3)
> plot.gof.censored.summary(sw.list.lnorm, digits=3)
> detach()
```

Note: You can also plot the results of goodness-of-fit tests by using the generic `plot` function. For example, typing `plot(sw.list.norm)` will bring up a menu of options for different kinds of plots, including the summary plots.

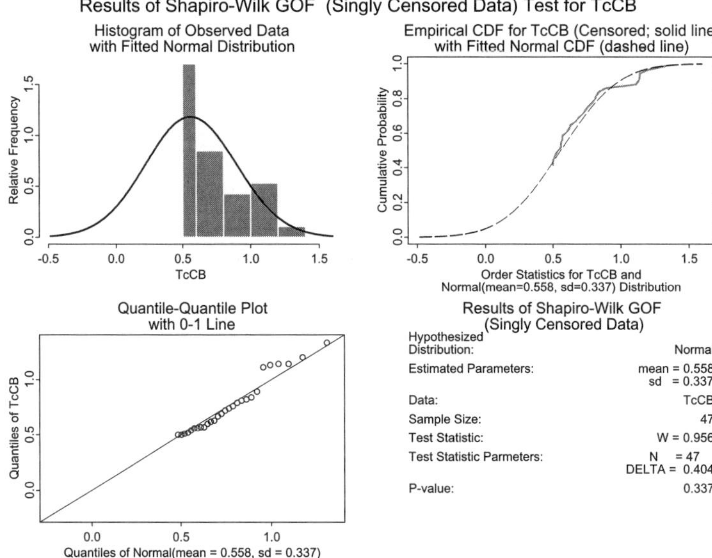

Figure 8.8. Companion plots for the Shapiro-Wilk test for normality for the singly censored Reference area TcCB data.

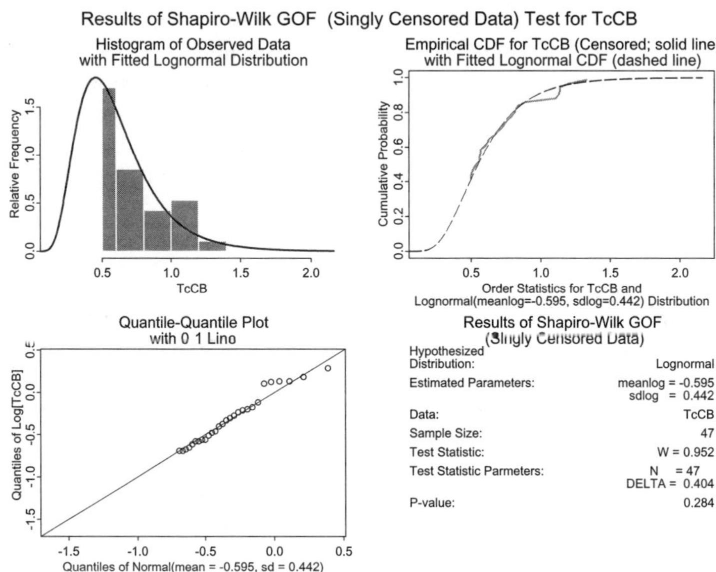

Figure 8.9. Companion plots for the Shapiro-Wilk test for lognormality for the singly censored Reference area TcCB data.

8.8.2   Nonparametric Tests to Compare Two Groups

In Chapter 7 we discussed various hypothesis tests to compare locations (central tendency) between two groups, including the Wilcoxon rank sum test, other linear rank tests, and the quantile test. In the presence of censored observations, you can still use a linear rank test (e.g., the Wilcoxon rank sum test) or quantile test as long as there are enough uncensored observations that can be ordered in a logical way. For example, if both samples are singly censored with the same censoring level and all uncensored observations are greater than the censoring level, then all censored observations receive the lowest ranks and are considered tied observations. Actually you can use linear rank tests even with multiply censored data as well.

We stated in Chapter 7 that the Wilcoxon rank sum test is a particular kind of linear rank test. Several authors have proposed extensions of the Wilcoxon rank sum test to the case of singly or multiply censored data, mainly in the context of survival analysis (e.g., Breslow, 1970; Cox, 1972; Gehan, 1965; Mantel, 1966; Peto and Peto, 1972; Prentice, 1978). Prentice (1978) showed how all of these proposed tests are extensions of a linear rank test to the case of censored observations. As for the case of complete data, different linear rank tests use different score functions, and some may be better than others at detecting a small shift in location, depending upon the true underlying distribution.

Prentice and Marek (1979), Latta (1981), and Millard and Deverel (1988) studied the behavior of several linear rank tests for censored data. For details, see the help file for two.sample.linear.rank.test.censored.

In Section 8.4.3 we compared the empirical cumulative distribution functions of copper concentrations in the alluvial fan and basin trough zones (see Table 8.3 and Figure 8.5). The plot indicates the distributions of concentrations are fairly similar. The two-sample linear rank test based on normal scores and a hypergeometric variance yields a p-value of 0.2, indicating no significant difference.

Menu

To perform the two-sample linear rank test to compare copper concentrations between the two geologic zones using the ENVIRONMENTALSTATS for S-PLUS pull-down menu, follow these steps.

1. Find **millard.deverel.88.df** in the Object Explorer.
2. On the S-PLUS menu bar, make the following menu choices: **EnvironmentalStats>Censored Data>Hypothesis Tests>Compare Two Samples**. This will bring up the Compare Two Samples for Censored Data dialog box.
3. The Data Set box should display **millard.deveral.88.silver.df**. For Variable 1 choose **Cu**, for Censor Ind. 1 choose **Cu.censored**, for Variable 2 choose **Zone**, and **check** the Variable 2 is a Grouping Variable box. For test select **normal.scores.2** and for variance select **hypergeometric**, then click **OK** or **Apply**.

Command

To perform the two-sample linear rank test to compare copper concentrations, type these commands.

```
> attach(millard.deverel.88.df)
> Cu.AF <- Cu[Zone=="Alluvial.Fan"]
> Cu.AF.cen <- Cu.censored[Zone=="Alluvial.Fan"]
> Cu.BT <- Cu[Zone=="Basin.Trough"]
> Cu.BT.cen <- Cu.censored[Zone=="Basin.Trough"]
> two.sample.linear.rank.test.censored(Cu.AF, Cu.AF.cen,
      Cu.BT, Cu.BT.cen, test="normal.scores.2",
      var="hypergeometric")
> detach()
```

8.9 Summary

- A sample of data contains ***censored observations*** if some of the observations are reported only as being below or above some censoring level.
- Environmental data often contain values reported as "less-than-detection-limit".
- Table 8.1 lists menu items and functions available in ENVIRONMENTALSTATS for S-PLUS for analyzing censored data.

9

Monte Carlo Simulation and Risk Assessment

9.1 Introduction

In the first 8 chapters of this book we have discussed several statistical tools for looking at data, modeling probability distributions, estimating distribution parameters and quantiles, constructing prediction and tolerance intervals, and comparing two or more groups. All of our examples have concentrated on assessing how much chemical is in the environment and comparing chemical concentrations to "background." But given that chemicals are in the environment, what happens when people or other living organisms are exposed to these chemicals?

Of course, "chemicals" in the environment are part of our everyday lives: they are in the food we eat, the water we drink, the air we breathe. Some are natural and others are synthetic, having been added either on purpose or as a by-product of a manufacturing process. There is no doubt that the chemical revolution of the past century has improved our lives immensely. But we have also learned that some chemicals that have improved some facet of our lives can also have devastating consequences upon our health and environment.

Several government agencies are charged with evaluating the potential health and ecological effects of environmental toxicants. Based on their assessments, these agencies set standards for acceptable concentration levels of these toxicants in air, water, soil, food, etc. The process of modeling exposure to a toxicant and predicting health or ecological effects is termed *risk assessment*.

Risk assessment is a process where science, politics, and psychology all intersect. Not surprisingly, it is also a field full of controversy. References discussing risk assessment and the concept of risk include Cothern (1996), Everitt (1999), Hallenbeck (1993), Laudan (1997), Lewis (1990), Lundgren and McMakin (1998), Molak (1997), Neely (1994), Rodricks (1992), Shaffner (1999), Suter (1993), and Walsh (1996). In this chapter, we will introduce basic mathematical models used in risk assessment and talk about how to use ENVIRONMENTALSTATS for S-PLUS to perform Monte Carlo simulation and probabilistic risk assessment. See Chapter 13 of Millard and Neerchal (2001) for a more in-depth discussion of Monte Carlo simulation and risk assessment.

9.2   Overview

Human and ecological risk assessment involves characterizing the exposure to a toxicant for one or several populations, quantifying the relationship between exposure (dose) and health or ecological effects (response), determining the risk (probability) of a health or ecological effect given the observed level(s) of exposure, and characterizing the uncertainty associated with the estimated risk (Hallenbeck, 1993, p. 1). Risk assessment is an enormous field of research and practice. The merits and disadvantages of particular exposure models (including fate and transport models) and dose-response models are not discussed in this chapter. Several journals, textbooks, and web sites are dedicated to risk assessment. Three standard texts include Hallenbeck (1993), Suter (1993), and Rodricks (1992). Other references are listed in the ENVIRONMENTALSTATS for S-PLUS help file *References: Risk Assessment.*

In the past, estimates of risk were often based solely on setting values of the input variables (e.g., body weight, dose, etc.) to particular point estimates and producing a single point estimate of risk, with little, if any, quantification of the uncertainty associated with the estimated risk. More recently, several practitioners have advocated "probabilistic" risk assessment, in which the input variables are considered random variables, so the result of the risk assessment is a probability distribution for predicted risk or exposure.

Usually, the equation describing risk or exposure is so complicated that it is not feasible to determine the output distribution using analytical methods, so the distribution of risk or exposure is derived via Monte Carlo simulation. This chapter discusses the concepts of Monte Carlo simulation, sensitivity and uncertainty analysis, and risk assessment, and shows you how to use S-PLUS and ENVIRONMENTALSTATS for S-PLUS to perform probabilistic risk assessment.

9.3   Monte Carlo Simulation

Monte Carlo simulation is a method of investigating the distribution of a random variable by simulating random numbers (Gentle, 1985). Usually, the random variable of interest, say Y, is some function of one or more other random variables:

$$Y = h(\underline{X}) = h\left(X_1, X_2, \ldots, X_k\right) \tag{9.1}$$

For example, Y may be an estimate of the median of a population with a Cauchy distribution, in which case the vector of random variables $\underline{X}$ represents k independent and identically distributed observations from some particular Cauchy distribution. As another example, Y may be the incremental lifetime cancer risk due to ingestion of soil contaminated with benzene (Thompson et al., 1992; Hamed and Bedient, 1997). In this case the random vector $\underline{X}$ may represent observations from several kinds of distributions that characterize exposure and dose-response, such as benzene concentration in the soil, soil ingestion rate, average body weight, the cancer potency factor for benzene, etc. These distributions may or may not be assumed to be independent of one another (Smith et al., 1992; Bukowski et al., 1995).

Sometimes the input variables X_1, X_2, ..., X_k are called ***input parameters***. This terminology can be confusing, however, since the input variables are often random variables and therefore have ***distribution parameters*** associated with their probability distributions (e.g., mean and standard deviation for a normally distributed input variable).

Sometimes the distribution of Y in Equation (9.1) can be derived analytically based on statistical theory (Springer, 1979; Slob, 1994). Often, however, the function h is complicated and/or the elements of the random vector $\underline{X}$ involve several kinds of probability distributions, making it difficult or impossible to derive the exact distribution of Y. In this case, Monte Carlo simulation can be used to approximate the distribution of Y. Monte Carlo simulation is often used in risk assessment, specifically in sensitivity and uncertainty analysis.

Monte Carlo simulation involves creating a large number of realizations of the random vector $\underline{X}$, say n, and computing Y for each of the n realizations of $\underline{X}$. The resulting distribution of Y, or some characteristic of this distribution (e.g., the mean), is then assumed to be "close" to the true distribution or distribution characteristic of Y. The adequacy of the approximation depends on a number of factors, including how well the mathematical relationship described in Equation (9,1) reflects the true relationship between Y and $\underline{X}$, how well the specified distribution of $\underline{X}$ reflects its true distribution (including any possible dependencies between the individual elements of $\underline{X}$), and how many Monte Carlo samples or trials (n) are created.

Usually, Monte Carlo simulation involves generating random numbers from some specified theoretical probability distribution, such as a normal, lognormal, beta, etc. When the simulation is done based on an empirical distribution, this is also called bootstrapping (Efron and Tibshirani, 1993).

Vose (1996, p. 40) claims that the term "Monte Carlo" comes from the code name of an American project on the atom bomb during World War II, and not from the town of the same name in Monaco that is well known for its casinos. This, however, begs the question of the origin of the code name. Gentle (1985, p. 612) notes that Monte Carlo simulation was used extensively during World War II and immediately after during work on the development of the atomic bomb. He states that the name "Monte Carlo" dates from that time period and comes from the casino in Monaco with the same name. Hayes (1993, p. 114) states that the modern form of Monte Carlo simulation was invented at Los Alamos shortly after World War II by Stanislaw Ulam while he was working on a problem of the diffusion of neutrons, and that the term Monte Carlo method is "named in honor of the well-known generator of random integers (between 0 and 36) in the Mediterranean principality." References that address the issues of how to properly perform and report the results of a Monte Carlo simulation study include Hoaglin and Andrews (1975), Law and Kelton (1991), Burmaster and Anderson (1994), and Vose (1996).

9.3.1   Simulating the Distribution of the Sum of Two Normal Random Variables

Suppose X_1 and X_2 are two independent standard normal random variables. Then the distribution of

$$Y = h(X_1, X_2) = X_1 + X_2 \qquad (9.2)$$

is normal with a mean of 0 and a variance of 2. Suppose, however, that we do not know how to derive the distribution of Y. We can use Monte Carlo simulation to investigate the shape of the distribution of Y, as well as compute characteristics of the distribution (e.g., mean, median, standard deviation, quantiles, etc.)

Figure 9.1 displays the empirical and true distribution of Y, where the empirical distribution of Y was derived by using Monte Carlo simulation with 1,000 trials. That is, for each trial, two random numbers from a standard normal distribution were generated and added together. Figure 9.2 displays the empirical distribution based on 10,000 trials along with the true distribution. Table 9.1 displays some summary statistics for the two empirical distributions and compares them with the true population values. As we increase the number of Monte Carlo trials, the simulated distribution tends to get "closer" to the true distribution. This is called the ***Law of Large Numbers***.

Parameter	Empirical (1,000)	Empirical (10,000)	Population N(0, 2)
Mean	0.05	0.00	0
Standard Deviation	1.42	1.41	1.41
5th Percentile	-2.37	-2.33	-2.33
95th Percentile	2.42	2.31	2.33

Table 9.1. Comparison of empirical and population summary statistics.

Menu

To generate the empirical distribution of the sum of two independent standard normal random variables based on 1,000 Monte Carlo trials using the ENVIRONMENTALSTATS for S-PLUS pull-down menu follow these steps.

1.   On the S-PLUS menu bar, make the following menu choices: **EnvironmentalStats>Probability Distributions and Random Numbers>Random Numbers>Multivariate>Normal**. This will bring up the Multivariate Normal Random Numbers dialog box.
2.   For Sample Size type **1000**, for Set Seed with type **20**, for Number of Vars select **2, uncheck** the Print Results box, and for Save As type **rmvnorm.df**. Click **OK**.
3.   Find **rmvnorm.df** in the Object Explorer.

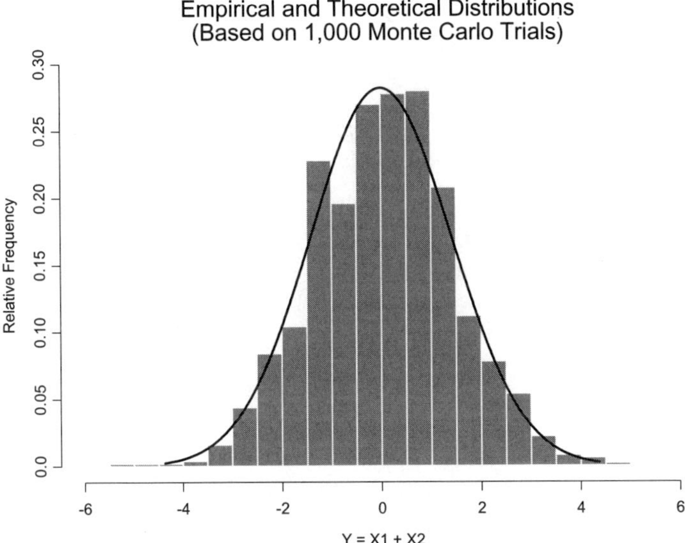

Figure 9.1. Empirical and theoretical distribution of the sum of two independent $N(0,1)$ random variables based on 1,000 Monte Carlo trials.

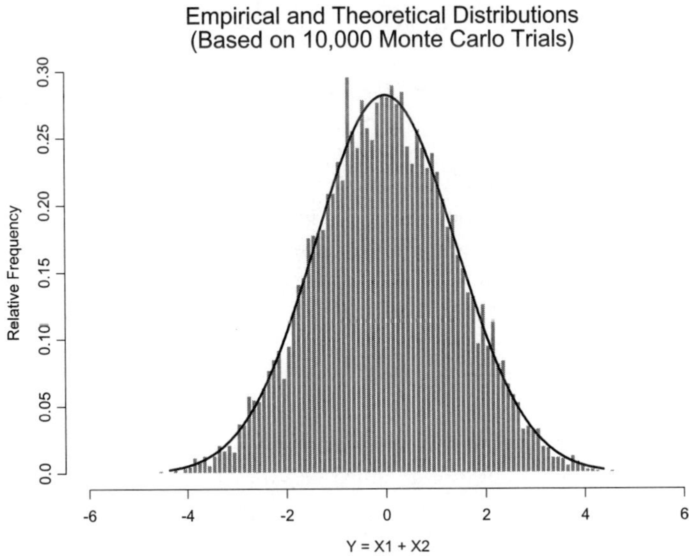

Figure 9.2. Empirical and theoretical distribution of the sum of two independent $N(0,1)$ random variables based on 10,000 Monte Carlo trials.

4. On the S-PLUS menu bar, make the following menu choices: **Data>Transform**. This will bring up the Transform dialog box.
5. For Data Set select **rmvnorm.df**, for Target Column type **X.1.plus.X.2**, in the Expression box type **X.1 + X.2**, and click **OK** or **Apply**.

To produce the empirical distribution based on 10,000 Monte Carlo trials, in Step 2 above for Sample Size type **10000**.

Command

To generate the empirical distribution of the sum of two independent standard normal random variables based on 1,000 Monte Carlo trials type these commands.

```
> set.seed(20)
> rmvnorm.mat <-rmvnorm(n=1000)
> y <- rowSums(rmvnorm.mat)
```

or you can use the ENVIRONMENTALSTATS for S-PLUS functions `simulate.mv.matrix` and `eval.expr`:

```
> rmvnorm.mat <-simulate.mv.matrix(n=1000, seed=20)
> y <- eval.expr(Var.1+Var.2, data=rmvnorm.mat)
```

To produce the empirical distribution based on 10,000 Monte Carlo trials, set `n=10000` in the call to `rmvnorm` or `simulate.mv.matrix`.

9.4   Generating Random Numbers

A random number is a realization of a random variable, say X. For many people, the term random number initially conjures up an image of somehow choosing an integer between a specified lower and upper bound (e.g., 1 and 10), where each number is equally likely to be chosen. In that case, the random variable X is a discrete uniform random variable with probability density (mass) function given by:

$$f(x) = \Pr(X = x) = \frac{1}{10} \; ; \; x = 1, 2, \ldots, 10 \qquad (9.3)$$

In general, a random number can be a realization of a random variable from any kind of probability distribution (e.g., uniform, normal, lognormal, gamma, empirical, etc.)

9.4.1   Generating Random Numbers from a Uniform Distribution

The S-PLUS function `runif` generates ***pseudo-random numbers*** from a (continuous) uniform distribution. Uniform random number generation in S-PLUS is adapted from Marsaglia et al. (1973), which couples a multiplicative-congruential generator with a feedback shift register. References that discuss generating pseudo-random numbers include Kennedy and Gentle (1980), Rubenstein (1981), Ripley (1987), Law and Kelton (1991), Hayes (1993), and Barry (1996). Venables and Ripley (1999, pp. 116–117) explain the S-PLUS pseudo-random number generator in detail.

Pseudo-random number generators start with an initial *seed*, and then appear to generate random numbers, although these numbers are actually generated by a deterministic mechanism. Each time a set of random numbers is generated, the value of the seed changes. If you start with the same seed, you will get the same sequence of pseudo-random numbers. You can use the S-PLUS function set.seed to set the seed of the random number generator. Also, S-PLUS and ENVIRONMENTALSTATS for S-PLUS dialog boxes for generating random numbers allow you to set the seed.

The *period* of a random number generator is the number of random numbers that can be generated before the sequence repeats itself. For most starting seeds, the period of the random number generator in S-PLUS is $2^{30} \times 4{,}292{,}868{,}097 \approx 4.6 \times 10^{18}$ (about 4.6 quintillion) (Venables and Ripley, 1999, p. 117).

9.4.2 Generating Random Numbers from an Arbitrary Distribution

As we saw in Chapter 4, the S-PLUS and ENVIRONMENTALSTATS for S-PLUS functions of the form `rabb` (where `abb` denotes the abbreviation of the distribution) generate random numbers from several theoretical probability distributions. For example, the function `rnorm` generates random numbers from a normal distribution. You can generate random numbers from the S-PLUS menu by selecting **Data>Random Numbers**, and you can generate random numbers from the ENVIRONMENTALSTATS for S-PLUS menu by selecting **EnvironmentalStats>Probability Distributions and Random Numbers>Random Numbers**.

To generate random numbers from a specified probability distribution, most computer software programs use the inverse transformation method (Rubenstein, 1981, pp. 38–43; Law and Kelton, 1991, pp. 465–474; Vose, 1996, pp. 39–40). Suppose the random variable U has a U[0,1] distribution, that is, a uniform distribution over the interval [0,1]. Let F_X denote the cumulative distribution function (cdf) of the specified probability distribution. Then the random variable X defined by:

$$X = F_X^{-1}(U) \tag{9.4}$$

has the specified distribution, where the quantity F_X^{-1} denotes the inverse of the cdf function F_X. Thus, to generate a set of random numbers from any distribution, all you need is a set of random numbers from a U[0,1] distribution and a function that computes the inverse of the cdf function for the specified distribution. Figure 9.3 illustrates the inverse transformation method for a standard normal distribution, with $U = 0.8$. In this case, the random number generated is $\Phi^{-1}(0.8) = 0.8416212$.

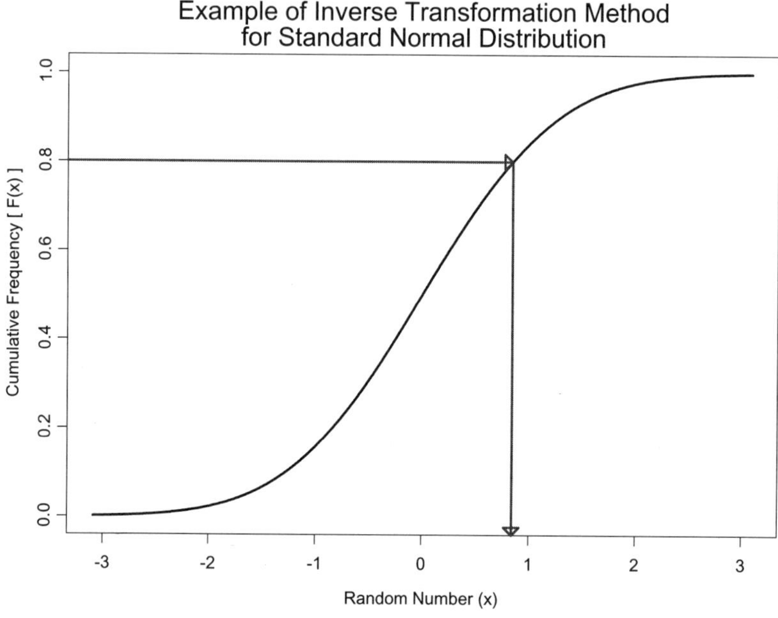

Figure 9.3. Example of the inverse transformation method of generating random numbers from a specified distribution.

9.4.3   Latin Hypercube Sampling

Latin Hypercube sampling, sometimes abbreviated LHS, is a method of sampling from a probability distribution (one random variable) or a joint probability distribution (several random variables) that ensures all portions of the probability distribution are represented in the sample. It was introduced in the published literature by McKay et al. (1979). Other references include Iman and Conover (1980, 1982), and Vose (1996, pp. 42–45). Latin Hypercube sampling is an extension of quota sampling, and when applied to the joint distribution of k random variables, can be viewed as a k-dimensional extension of Latin square sampling, thus the name (McKay et al., 1979, p. 240).

Latin Hypercube sampling was introduced to overcome the following problem in Monte Carlo simulation based on simple random sampling (SRS). Suppose we want to generate random numbers from a specified distribution. If we use simple random sampling, there is a low probability of getting very many observations in an area of low probability of the distribution. For example, if we generate n observations from the distribution, the probability that none of these observations falls into the upper 98th percentile of the distribution is 0.98^n. So, for example, there is a 13% chance that out of 100 random numbers, none will fall at or above the 98th percentile. If we are interested in reproducing the shape of the distribution, we will need a very large number of observations to ensure that we can adequately characterize the tails of the distribution (Vose, 1996, p. 40).

Latin Hypercube sampling was developed in the context of using computer models that required enormous amounts of time to run and for which only a limited number of Monte Carlo simulations could be implemented. In cases where it is fairly easy to generate tens of thousands of Monte Carlo trials, Latin Hypercube sampling may not offer any real advantage.

Latin Hypercube sampling works as follows for a single probability distribution. If we want to generate n random numbers from the distribution, the distribution is divided into n intervals of equal probability $1/n$. A random number is then generated from each of these intervals. For k independent probability distributions, LHS is applied to each distribution, and the resulting random numbers are matched at random to produce n random vectors of dimension k.

Figures 9.4 – 9.6 illustrate Latin Hypercube sampling for a sample size of n = 4, assuming a standard normal distribution. Figure 9.4 shows the four equal-probable intervals for a standard normal distribution in terms of the probability density function, and Figure 9.5 shows the same thing in terms of the cumulative distribution function. Figure 9.6 shows how Latin Hypercube sampling is accomplished using the inverse transformation method for generating random numbers. In this case, the interval [0,1] is divided into the four intervals [0, 0.25], [0.25, 0.5], [0.5, 0.75], and [0.75, 1]. Next, a uniform random number is generated within each of these intervals. For this example, the four numbers generated are (to two decimal places) 0.04, 0.35, 0.70, and 0.79. Finally, the standard normal random numbers associated with the inverse cumulative distribution function of the four uniform random numbers are computed.

9.4.4 Example of Simple Random Sampling versus Latin Hypercube Sampling

Figure 9.7 displays a histogram of 50 observations based on a simple random sample from a standard normal distribution. Figure 9.8 displays the same thing based on a Latin Hypercube sample. You can see that the form of the histogram constructed with the Latin Hypercube sample more closely resembles the true underlying distribution.

Menu

To generate the simple random sample used to create the histogram in Figure 9.7 using the ENVIRONMENTALSTATS for S-PLUS pull-down menu, follow these steps.

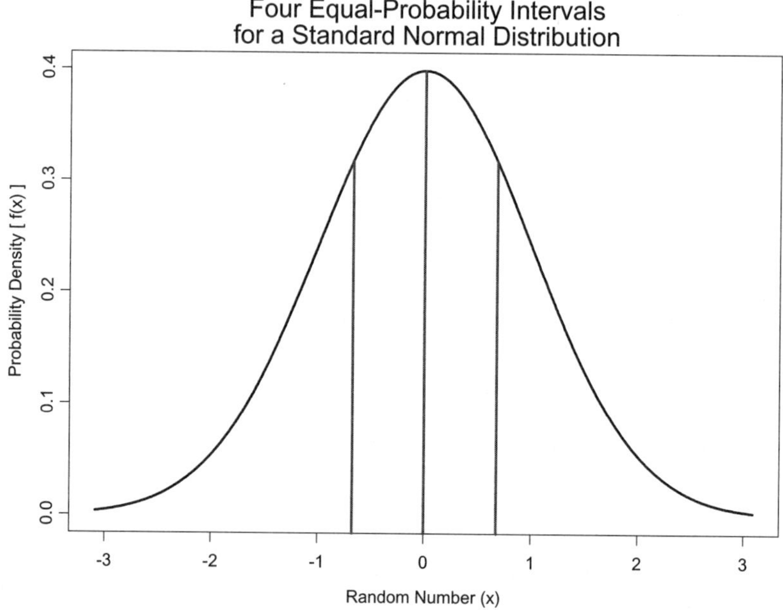

Figure 9.4. Four equal-probability intervals for a N(0, 1) distribution.

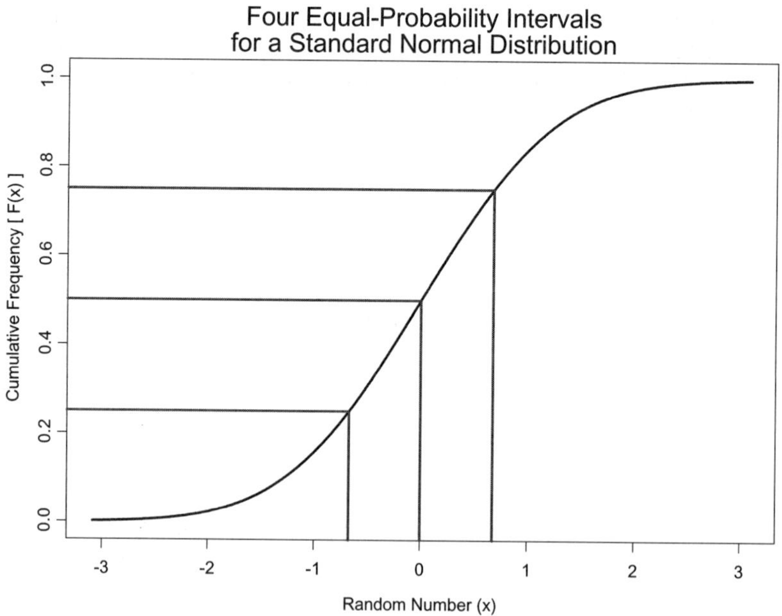

Figure 9.5. Four equal-probability intervals for a N(0, 1) distribution.

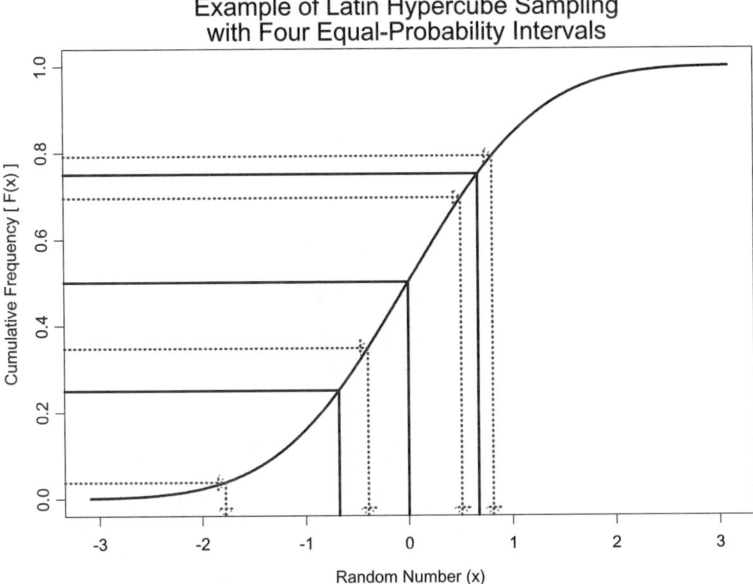

Figure 9.6. Visual explanation of generating four random numbers from a N(0, 1) distribution using Latin Hypercube sampling.

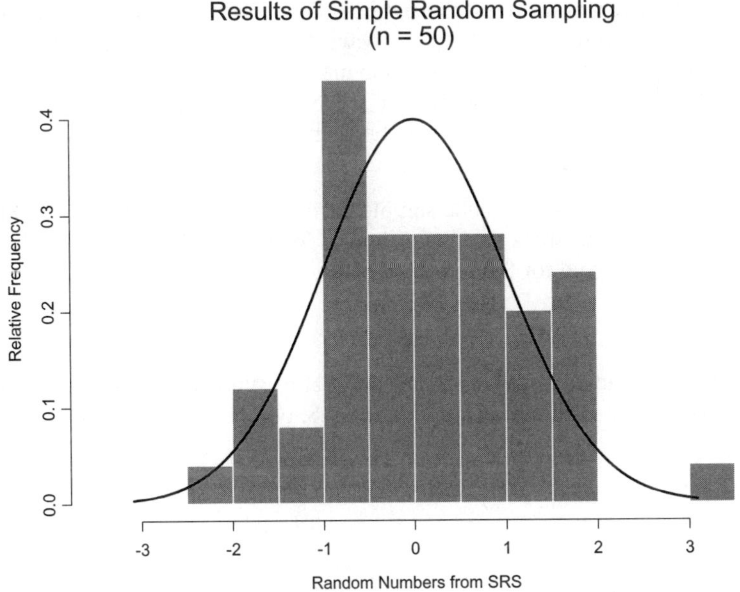

Figure 9.7. Results of simple random sampling from a N(0, 1) distribution.

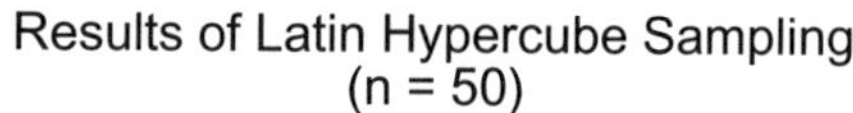

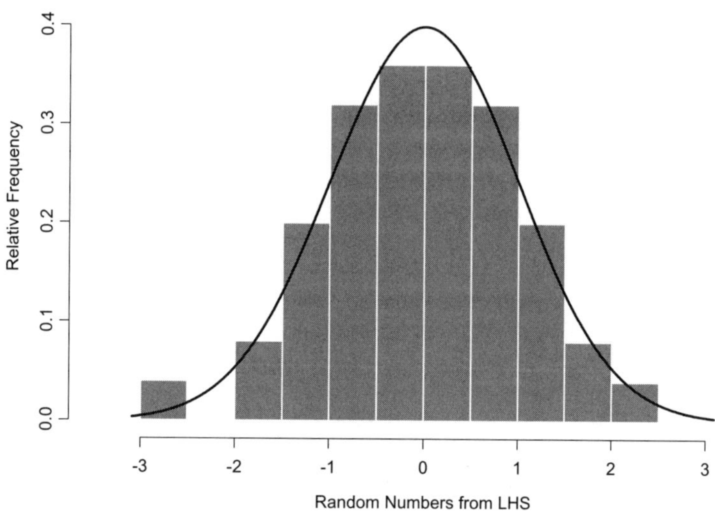

Figure 9.8. Results of Latin Hypercube sampling from a N(0, 1) distribution.

1. On the S-PLUS menu bar, make the following menu choices: **EnvironmentalStats>Probability Distributions and Random Numbers>Random Numbers>Univariate**. This will bring up the Univariate Random Number Generation dialog box.
2. For Sample Size type **50**, for Set Seed with type **209**, for Sampling Method select **Random**, **uncheck** the Print Results box, and for Save As type **x.srs**.

To generate the Latin Hypercube sample used to create the histogram in Figure 9.8 follow the same steps as above, but in Step 2 for Sample Method select **Latin Hypercube** and for Save As type **x.lhs**.

Command

To generate the simple random sample and Latin Hypercube sample used to create the histograms in Figure 9.7 and Figure 9.8, type these commands.

```
> x.srs <- simulate.vector(50, seed=209)
> x.lhs <- simulate.vector(50, sample.method="LHS",
    seed=209)
```

9.4.5 Properties of Latin Hypercube Sampling

Let Y denote the outcome variable for one trial of a Monte Carlo simulation, and suppose Y is a function of k independent random variables as shown in Equation (9.1) above. McKay et al. (1979) consider the class of estimators of the form

$$T = T\left(\underline{Y}\right) = T\left(Y_1, Y_2, \ldots, Y_n\right) = \frac{1}{n}\sum_{i=1}^{n} g\left(Y_i\right) \tag{9.5}$$

where g is an arbitrary function. This class of estimators includes the mean, the rth sample moment, and the empirical cumulative distribution function. Setting

$$\tau = E\left[T\left(\underline{Y}\right)\right] \tag{9.6}$$

McKay et al. (1979) show that under LHS, T is an unbiased estimator of τ, and also, if h in Equation (9.1) is monotonic in each of its arguments and g is monotonic, then the variance of T under LHS is less than or equal to the variance of T under SRS.

Stein (1987) shows that the variance of the sample mean of Y under LHS is asymptotically less than the variance of the sample mean under simple random sampling whether or not the function h is monotonic in its arguments. Unfortunately, for most cases of LHS, the formula for the true variance of the sample mean is difficult to derive, and thus a good estimate of true variance is not available. Using the usual formula of dividing the sample variance by the sample size will usually overestimate the true variance of the sample mean.

Iman and Conover (1980) and Stein (1987) suggest producing several independent Latin Hypercube samples, say N, and for each Latin Hypercube sample computing the sample mean based on the n observations within that sample. (They call this method *replicated Latin Hypercube sampling*.) You can then estimate the variance of the sample mean by computing the usual sample variance of these N sample means. Note that this method can be applied to any quantity of interest, such as the median, 95th percentile, etc.

9.5 Uncertainty and Sensitivity Analysis

Uncertainty analysis and sensitivity analysis are terms used to describe various methods of characterizing the behavior of a complex mathematical/computer model. The model in Equation (9.1) above is different from most conventional statistical models, where the form of the model is:

$$Y = h\left(\underline{X}\right) + \varepsilon \tag{9.7}$$

(e.g., linear regression models, generalized linear models, nonlinear regression models, etc.). In Equation (9.7), the vector $\underline{X}$ is assumed to be set or observed at fixed values, and for fixed values of $\underline{X}$ the response variable Y deviates about its mean value according to the distribution of the error term ε. This kind of model is useful when we are interested in the specific relationship between Y and $\underline{X}$, and we want to predict the value of Y for a specified value of $\underline{X}$. Furthermore, this kind of model is fit using paired observations of Y and $\underline{X}$.

In Equation (9.1), Y is assumed to be observed without error, that is, the value of Y is deterministic for a set value of $\underline{X}$. The output variable Y, however, is a random variable because the input variables $X_1, X_2, \ldots, X_k$ are assumed to be a combination of random variables and constant terms. This kind of model is useful when we are interested in describing the distribution of Y taken over all possible (read as "reasonable and realistic") combinations of the input variables. Furthermore, paired observations of Y and $\underline{X}$ are usually not available to validate this kind of model, hence, there is some amount of unquantifiable uncertainty associated with this kind of model.

Uncertainty analysis involves describing the variability or distribution of values of the output variable Y that is due to the collective variation in the input variables $\underline{X}$ (Iman and Helton, 1988, p. 72). This description usually involves graphical displays such as histograms, empirical density plots, and empirical cdf plots, as well as summary statistics such as the mean, median, standard deviation, coefficient of variation, 95th percentile, etc.

Sensitivity analysis involves determining how the distribution of Y changes with changes in the individual input variables. It is used to identify which input variables contribute the most to the variation or uncertainty in the output variable Y (Iman and Helton, 1988, p. 72). Sensitivity analysis is also used in a broader sense to determine how changing the distributions of the input variables and/or their assumed correlations or even changing the form of the model affects the output (Thompson et al., 1992; Smith et al., 1992; Cullen, 1994; Shlyakhter, 1994; Bukowski et al., 1995; Hamed and Bedient, 1997; USEPA, 1997).

9.5.1   Important versus Sensitive Parameters

Two useful concepts associated with sensitivity analysis are important parameters (variables) and sensitive parameters (variables) (Crick et al., 1987 as cited in Hamby, 1994, p. 137). ***Sensitive parameters*** have a substantial influence on the resulting distribution of the output variable Y, that is, small changes in the value of a sensitive parameter result in substantial changes in Y. ***Important parameters*** have some amount of uncertainty and/or variability associated with them and this variability contributes substantially to the resulting variability in the output variable Y.

Figure 9.9 below illustrate these two concepts for the simple case of three input variables. The top plot is an example of an important variable. An important variable is always sensitive. The middle plot is an example of a variable that is not sensitive, and hence not important. The bottom plot is an example of a variable that is not important. This variable may not be sensitive (like the one in the middle plot), or it may be sensitive like the one in the top plot but it is not important because of its limited variability.

X1 Important (Thus Sensitive)

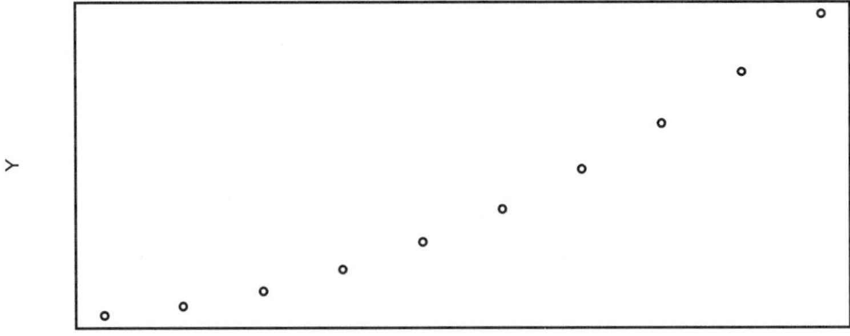

X2 Not Sensitive (Thus Not Important)

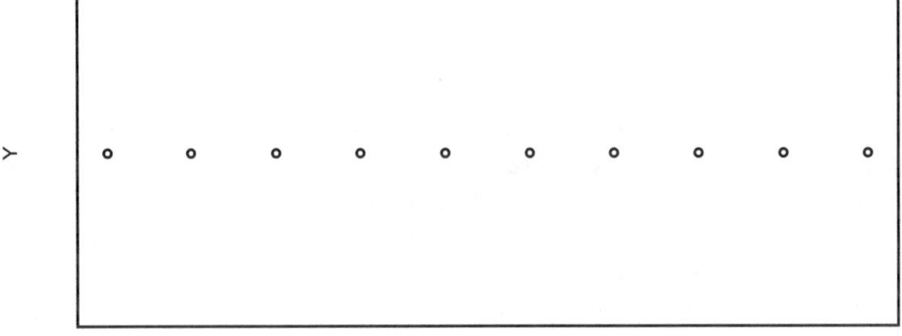

X3 Not Important

Figure 9.9. Three examples of the concepts of important and sensitive parameters.

9.5.2   Uncertainty versus Variability

The terms uncertainty and variability have specific meanings in the risk assessment literature that do not necessarily match their meanings in the statistical literature or everyday language. The term ***variability*** refers to the inherent heterogeneity of a particular variable (parameter). For example, there is natural variation in body weight and height between individuals in a given population. The term ***uncertainty*** refers to a lack of knowledge about specific parameters, models, or factors (Morgan and Henrion, 1990; Hattis and Burmaster, 1994; Rowe, 1994; Bogen, 1995; USEPA, 1997, p. 10). For example, we may be uncertain about the true distribution of exposure to a toxic chemical in a population (parameter uncertainty due to lack of data, measurement errors, sampling errors, systematic errors, etc.), or we may be uncertain how well our model of incremental lifetime cancer risk reflects reality (model uncertainty due to simplification of the process, misspecification of the model structure, model misuse, use of inappropriate surrogate variables, etc.), or we may be uncertain about whether a chemical is even present at a site of concern (scenario uncertainty due to descriptive errors, aggregation errors, errors in professional judgment, incomplete analysis, etc.).

We can usually reduce uncertainty through further measurement or study. We cannot reduce variability, since it is inherent in the variable. Note that in the risk assessment literature, measurement error contributes to uncertainty; we can decrease uncertainty by decreasing measurement error. In the statistical literature, measurement error is one component of the variance of a random variable. Note that a parameter (input variable) may have little or no variability associated with it, yet still have uncertainty associated with it (e.g., the speed of light is constant, but we only know its value to a given number of decimal places).

This meaning of the term "uncertainty" should not be confused with the term "uncertainty analysis" described above. Uncertainty analysis characterizes the distribution of the output variable Y. The output variable Y varies due to the fact that the input variables are random variables. The distributions of the input random variables reflect both variability (inherent heterogeneity) and uncertainty (lack of knowledge).

9.5.3   Sensitivity Analysis Methods

Sensitivity analysis methods can be classified into three groups: one-at-a-time deviations from a baseline case, factorial design and response surface modeling, and Monte Carlo simulation (Hamby, 1994). Each of these kinds of sensitivity analysis is briefly discussed below. For more detailed information, see the section Sensitivity Analysis Methods in the help file *Glossary: Uncertainty and Sensitivity Analysis* in ENVIRONMENTALSTATS for S-PLUS. Several studies indicate that using Monte Carlo simulation in conjunction with certain sensitivity measures usually provides the best method of determining sensitivity of the parameters.

One-at-a-Time Deviations from a Baseline Case

These sensitivity analysis methods include differential analysis and measures of change in output to change in input. Differential analysis is simply approximating the variance of the output variable Y at a particular value of the input vector $\underline{X}$ (called the **baseline case**) by using a first-order Taylor series expansion (Kotz and Johnson, 1985, Volume 8, pp. 646–647; Downing et al., 1985; Seiler, 1987; Iman and Helton, 1988; Hamby, 1994, pp. 138–139). This approximating equation for the variance of Y is useful for quantifying the proportion of variability in Y that is accounted for by each input variable. Unfortunately, the approximation is usually good only in a small region close to the baseline case, and the relative contribution of each input variable to the variance of Y may differ dramatically for differently chosen baseline cases. Also, differential analysis requires the calculation of partial derivatives, which may or may not be a simple task, depending on the complexity of the input function h in Equation (9.1).

Measures of change in output to change in input include the ratio of percent change in Y to percent change in X_i (Hamby, 1994, pp. 139–140), the ratio of percent change in Y to change in X_i in units of the standard deviation of X_i (Hamby, 1994, p. 140; Finley and Paustenbach, 1994), the percent change in Y as X_i ranges from its minimum to maximum value (Hamby, 1994, p. 140), and spider plots, which are plots of Y versus percent change in X_i, or Y versus percentiles of X_i (Vose, 1996, pp. 282–283).

Factorial Design and Response Surface Modeling

The concepts of factorial designs and response surfaces come from the field of experimental design (Box et al., 1978). In the context of sensitivity analysis for computer models, n distinct values of the input vector $\underline{X}$ are chosen (usually reflecting the possible ranges and medians of each of the k input variables), and the model output Y is recorded for each of these input values. Then a multiple linear regression model (a **response surface**) is fit to these data. The fitted model is called the fitted response surface, and this response surface is used as a replacement for the computer model (Downing et al., 1985; Iman and Helton, 1988; Hamby, 1994, p. 140). The sensitivity analysis is based on the fitted response surface. The reason for using a response surface to replace the actual model output is that some computer models are, or used to be, very costly to run, whereas computing output based on a response surface is relatively inexpensive.

One way to rank the importance of the variables in the response surface is to simply compare the magnitudes of the estimated coefficients. The estimated coefficients, however, depend on the units of the predictor variables in the model, so most analysts use standardized regression coefficients (Iman and Helton, 1988; Hamby, 1994, p. 146). The standardized regression coefficients are simply the coefficients that are obtained from fitting the response surface model based on the "standardized" output variable and the "standardized" predictor (input) variables. That is, for each variable, each observation is replaced by subtracting the mean (for that variable) from the observation and dividing by the standard deviation (for that variable).

Weisberg (1985, p. 186) warns against using standardized coefficients to determine the importance of predictor variables because they depend on the range of the predictor variables. So, for example, the variable X_3 in Figure 9.9 above may be very important but it has a very limited range. In the context of sensitivity analysis for computer models, however, if the analyst is comfortable with assuming the range of input variables will not change appreciably in the future, then standardized coefficients are an acceptable way to rank the importance of variables.

Iman and Helton (1988) compared uncertainty and sensitivity analysis of several models based on differential analysis, factorial design with a response surface model, and Monte Carlo simulation using Latin Hypercube sampling. The main outcome they looked at for uncertainty analysis was estimating the cumulative distribution function. They found that the models were too mathematically complex to be adequately represented by a response surface. Also, the results of differential analysis gave widely varying results depending on the values chosen for the baseline case. The method based on Monte Carlo simulation gave the best results.

Monte Carlo Simulation

Monte Carlo simulation is used to produce a distribution of Y values based on generating a large number of values of the input vector $\underline{X}$ according to the joint distribution of $\underline{X}$. There are several possible ways to produce a distribution for Y, including varying all of the input parameters (variables) simultaneously, varying one parameter at a time while keeping the others fixed at baseline values, or varying the parameters in one group while keeping the parameters in the other groups at fixed baseline values. Sensitivity methods that can be used with Monte Carlo simulation results include the following:

- **Histograms, Empirical CDF Plots, Percentiles of Output**. A simple graphical way to assess the effect of different input variables or groups of input variables on the distribution of Y is to look at how the histogram and empirical cdf of Y change as you vary one parameter at a time or vary parameters by groups (Thompson et al., 1992). Various quantities such as the mean, median, 95th percentile, etc. can be displayed on these plots as well.
- **Scatterplots**. Another simple graphical way to assess the effect of different input variables on the distribution of Y and their relationship to each other is to look at pair-wise scatterplots.
- **Correlations and Partial Correlations**. A quantitative measure of the relationship between Y and an individual input variable X_i is the correlation between these two variables, computed based on varying all parameters simultaneously (Saltelli and Marivoet, 1990; Hamby, 1994, pp. 143–144). Recall from Chapter 9 that the Pearson product moment correlation measures the strength of *linear* association between Y and X_i, while the Spearman rank and Kendall's tau correlation measures the strength of any *monotonic* relationship between Y and X_i. Vose (1996,

pp. 280–282) suggests using **tornado charts**, which are simply horizontal barcharts displaying the values of the correlations. Individual correlations are hard to interpret when some or most of the input variables are highly related to one another. One way to get around this problem is to look at partial correlation coefficients. See the section Sensitivity Analysis Methods in the ENVIRONMENTALSTATS for S-PLUS help file *Glossary: Uncertainty and Sensitivity Analysis Methods* for more information.

- **Change in Output to Change in Input**. Any of the types of measures that are described above under the subsection *One-at-a-Time Deviations from a Baseline Case* can be adapted to the results of a Monte Carlo simulation. Additional measures include relative deviation, in which you vary one parameter at a time and compute the coefficient of variation (CV) of Y for each case (Hamby, 1994, p. 142), and relative deviation ratio, in which you vary one parameter at a time and compute the ratio of the CV of Y to the CV of X_i (Hamby, 1994, pp. 142–143).

- **Response Surface**. This methodology that was described above in the subsection *Factorial Design and Response Surface Modeling* can be adapted to the results of a Monte Carlo simulation. In this case, use the model input and output to fit a regression equation (possibly stepwise) and then use standardized coefficients to rank the input variables (Iman and Helton, 1988; Saltelli and Marivoet, 1990).

- **Comparing Groupings within Input Distributions Based on Partitioning the Output Distribution**. One final method of sensitivity analysis that has been used with Monte Carlo simulation is to divide the distribution of the output variable Y into two or more groups, and then to compare the distributions of an input variable that has been split up based on these groupings (Saltelli and Marivoet, 1990; Hamby, 1994, pp. 146–150; Vose, 1996, pp. 296–298). For example, you could divide the distribution of an input variable X_i into two groups based on whether the values yielded a value of Y below the median of Y or above the median of Y. You could then use any of the Trellis functions to compare the distributions of these two groups, compare these distributions with a goodness-of-fit test, or compare the means or medians of these distributions with the t-test or Wilcoxon rank sum test. A significant difference between the two distributions is an indication that the input variable is important in determining the distribution of Y.

9.5.4   Uncertainty Analysis Methods

A specific model such as Equation (9.1) with specific joint distributions of the input variables leads to a specific distribution of the output variable. The process of describing the distribution of the output variable is called **uncertainty analysis** (Iman and Helton, 1988, p. 72). This description usually involves graphical displays such as histograms, empirical distribution plots, and empirical cdf plots, as well as summary statistics such as the mean, median, standard deviation, coefficient of variation, 95th percentile, etc.

Sometimes the distribution of Y in Equation (9.1) can be derived analytically based on statistical theory (Springer, 1979; Slob, 1994). For example, if the function h describes a combination of products and ratios, and all of the input variables have a lognormal distribution, then the output variable Y has a lognormal distribution as well, since products and ratios of lognormal random variables have lognormal distributions. Many risk models, however, include several kinds of distributions for the input variables, and some risk models are not easily described in a closed algebraic form. In these cases, the exact distribution of Y can be difficult or almost impossible to derive analytically.

The rest of this section briefly describes some methods of uncertainty analysis based on Monte Carlo simulation. For more information on uncertainty analysis, see the section Uncertainty Analysis Methods in the help file *Glossary: Uncertainty and Sensitivity Analysis* in ENVIRONMENTALSTATS for S-PLUS.

Quantifying Uncertainty with Monte Carlo Simulation

When the distribution of Y cannot be derived analytically, it can usually be estimated via Monte Carlo simulation. Given this simulated distribution, you can construct histograms or empirical density plots and empirical cdf plots, as well as compute summary statistics. You can also compute confidence bounds for specific quantities, such as percentiles. These confidence bounds are based on the assumption that the observed values of Y are randomly selected based on simple random sampling. When Latin Hypercube sampling is used to generate input variables and hence the output variable Y, the statistical theory for confidence bounds based on simple random sampling is not truly applicable (Easterling, 1986; Iman and Helton, 1991, p. 593; Stein, 1987). Most of the time, confidence bounds that assume simple random sampling but are applied to the results of Latin Hypercube sampling will probably be too wide.

Quantifying Uncertainty by Repeating the Monte Carlo Simulation

One way around the above problem with Latin Hypercube sampling is to use replicated Latin Hypercube sampling, that is, repeat the Monte Carlo simulation numerous times, say N, so that you have a collection of N empirical distributions of Y, where each empirical distribution is based on n observations of the input vector $\underline{X}$. You can then use these N replicate distributions to assess the variability of the sample mean, median, 95th percentile, empirical cdf, etc. Obviously, this is a computer intensive process.

A simpler process is to repeat the simulation just twice and compare the values of certain distribution characteristics, such as the mean, median, 5th, 10th, 90th and 95th percentiles, and also to graphically compare the two empirical cdf plots. If the values between the two simulations are within a small percentage of each other, then you can be fairly confident about characterizing the distribution of the output variable. Iman and Helton (1991) did this for a very complex risk assessment for a nuclear power plant and found a remarkable agreement in the empirical cdf's. Thompson et al. (1992) did the same thing for a risk assessment for incremental lifetime cancer risk due to ingestion or dermal contact with soil contaminated with benzene.

You may also want to compare the results of the original simulation with a simulation that uses say twice as many Monte Carlo trials (e.g., Thompson et al., 1992). Barry (1996) warns that if the moments of the simulated distribution do not appear to stabilize with an increasing number of Monte Carlo trials, this can mean that they do not exist. For most risk models, however, the true distributions of any random variables involved in a denominator in Equation (9.1) are bounded above 0, so the moments will exist. If a random variable involved in the denominator has a mean or median that is close to 0, it is important to use a bounded or truncated distribution to assure the random variable stays sufficiently far away from 0.

Quantifying Uncertainty Based on Mixture Distributions

To account for the uncertainty in specifying the distribution of the input variables, some authors suggest using mixture distributions to describe the distributions of the input variables (e.g., Hoffman and Hammonds, 1994; Burmaster and Wilson, 1996). For each input variable, a distribution is specified for the parameter(s) of the input variable's distribution. Some authors call random variables with this kind of distribution second-order random variables (e.g., Burmaster and Wilson, 1996).

For example we may assume the first input variable comes from a lognormal distribution with a certain mean and coefficient of variation (CV). The mean is unknown to a certain degree, and so we may specify that the mean comes from a uniform distribution with a given set of upper and lower bounds. We can also specify a distribution for the CV. In this case, the Monte Carlo simulation can be broken down into two stages. In the first stage, a set of parameters is generated for each input distribution. In the second stage, n realizations of the input vector $\underline{X}$ are generated based on this one set of distribution parameters. This two-stage process is repeated N times, so that you end up with N different empirical distributions of the output variable Y, and each empirical distribution is based on n observations of the input vector $\underline{X}$.

For a fixed number of Monte Carlo trials nN, the optimal combination of n and N will depend on how the distribution of the output variable Y changes relative to variability in the input distribution parameter(s) versus variability in the input variables themselves. For example, if the distribution of Y is very sensitive to changes in the input distribution parameters, then N should be large relative to n. On the other hand, if the distribution of Y is relatively insensitive to the values of the parameters of the input distributions, but varies substantially with the values of the input variables, then N may be small relative to n.

9.5.5   Caveat

An important point to remember is that no matter how complex the mathematical model in Equation (9.1) is, or how extensive the uncertainty and sensitivity analyses are, there is always the question of how well the mathematical model reflects reality. The only way to attempt to answer this question is with data collected directly on the input and output variables. Often, however, it is not possible to do this, which is why the model was constructed in the first place.

For example, in order to attempt to directly verify a model for incremental lifetime cancer risk for a particular exposed population within a particular geographical region, you have to collect data on lifetime exposure for each person and the actual proportion of people who developed that particular cancer within their lifetime, accounting for competing risks as well, and compare these data to similar data collected on a proper control population. A controlled experiment that involves exposing a random subset of a particular human population to a toxin and following the exposed and control group throughout their lifetimes for the purpose of a risk assessment is not possible to perform for several reasons, including ethical and practical ones. Rodricks (1992) is an excellent text that discusses the complexities of risk assessment based on animal bioassay and epidemiological studies.

9.6   Risk Assessment

This section discusses the concepts and practices involved in risk assessment, and gives examples of how to use ENVIRONMENTALSTATS for S-PLUS to perform probabilistic risk assessment. This information is also contained in the help file *Glossary: Risk Assessment*.

9.6.1   Definitions

It will be helpful to start by defining common terms and concepts used in risk assessment.

Risk

The common meaning of the term **risk** when used as a noun is "the chance of injury, damage, or loss." Thus, risk is a probability, since "chance" is another term for "probability."

Risk Assessment

Risk assessment is the practice of gathering and analyzing information in order to predict future risk. Risk assessment has been commonly used in the fields of insurance, engineering, and finance for quite some time. In the last couple of decades it has been increasingly applied to the problems of predicting human health and ecological effects from exposure to toxicants in the environment (e.g., Hallenbeck, 1993; Suter, 1993). In this chapter, the term risk assessment is applied in the context of human health and ecological risk assessment.

The basic model that is often used as the foundation for human health and ecological risk assessment is:

$$Risk = Dose \times \Pr\left(Effect \text{ per } Unit\,Dose\right) \qquad (9.8)$$

That is, the risk of injury (the *effect*) to an individual is equal to the amount of toxicant the individual absorbs (the *dose*) times the probability of the effect occurring for a single unit of the toxicant. If the effect is some form of cancer, the second term on the right-hand side of Equation (9.8) is often called the *cancer slope factor* (abbreviated CSF) or the *cancer potency factor* (abbreviated CPF).

The first term on the right-hand side of Equation (9.8), the dose, is estimated by identifying sources of the toxicant and quantifying their concentrations, identifying how these sources will expose an individual to the toxicant (via fate and transport models), quantifying the amount of exposure an individual will receive, and estimating how much toxicant the individual will absorb at various levels of exposure. Sometimes the dose represents the amount of toxicant absorbed over a lifetime, and sometimes it represents the amount absorbed over a shorter period of time.

The second term on the right-hand side of Equation (9.8), the probability of an effect (CSF or CPF), is estimated from a *dose-response curve*, a model that relates the probability of the effect to the dose received. Dose-response curves are developed from controlled laboratory experiments on animals or other organisms, and/or from epidemiological studies of human populations (Hallenbeck, 1993; Piegorsch and Bailer, 1997). Millard and Neerchal (2001, p. 586) fit a dose-response curve for the probability that a rat develops thyroid tumors as a function of dose of ethylene thiourea (ETU).

Risk assessment involves three major steps (Hallenbeck, 1993, p. 1; USEPA, 1995c):

- **Hazard Identification**. Describe the effects (if any) of the toxicant on laboratory animals, humans, and/or wildlife species, based on documented studies. Describe the quality and relevance of the data from these studies. Describe what is known about how the toxicant produces these effects. Describe the uncertainties and subjective choices or assumptions associated with determining the degree of hazard of the toxicant.

- **Dose-Response Assessment**. Describe what is known about the biological mechanism that causes the health or ecological effect. Describe what data, models, and extrapolations have been used to develop the dose-response curve for laboratory animals, humans, and/or wildlife species. Describe the routes and levels of exposure used in the studies to determine the dose-response curve, and compare them to the expected routes and levels of exposure in the population(s) of concern.

Describe the uncertainties and subjective choices or assumptions associated with characterizing the dose-response relationship.

- **Exposure Assessment**. Identify the sources of environmental exposure to the population(s) of concern. Describe what is known about the principal paths, patterns, and magnitudes of exposure. Determine average and "high end" levels of exposure. Describe the characteristics of the population(s) that is(are) potentially exposed. Determine how many members of the population are likely to be exposed. Describe the uncertainties and subjective choices or assumptions associated with characterizing the exposure for the population of concern.

Once these steps have been completed, some form of Equation (9.8) is usually used to estimate the risk for a particular population of concern. Both sensitivity analysis and uncertainty analysis should be applied to the risk assessment model to quantify the uncertainty associated with the estimated risk.

Risk Characterization

USEPA (1995c) distinguishes between the process of risk assessment and risk characterization. **Risk characterization** is the summarizing step of risk assessment that integrates all of the information from the risk assessment, including uncertainty and sensitivity analyses and a discussion of uncertainty versus variability, to form an overall conclusion about the risk. Risk assessment is the tool that a risk assessor uses to produce a risk characterization. A risk characterization is the product that is delivered to the risk assessor's client: the risk manager.

Risk Management

Risk management is the process of using information from risk characterizations (calculated risks), perceived risks, regulatory policies and statutes, and economic and social analyses in order to make and justify a decision (USEPA, 1995c). If the risk manager decides that the risk is not acceptable, he or she will order or recommend some sort of action to decrease the risk. If the risk manager decides that the risk poses minimal danger to the population of concern, he or she may recommend that no further action is needed at the present time.

Because the risk manager is the client of the risk assessor, the risk manager must be involved in the risk assessment process from the start, helping to determine the scope and endpoints of the risk assessment (USEPA, 1995c). Also, the risk manager must interact with the risk assessor throughout the risk assessment process, so that he or she may take responsibility for critical decisions. The risk manager, however, must be careful not to let non-scientific (e.g., political) issues influence the risk assessment. Non-scientific issues are dealt with at the risk management stage, not the risk assessment stage.

Risk Communication

USEPA (1995c) defines **risk communication** as exchanging information with the public. While the communication of risk from the risk assessor to the risk manager is accomplished through risk characterization, the communication of risk between the risk manager (or representatives of his or her agency) and the public is accomplished through risk communication. The risk characterization will probably include highly technical information, while risk communication should concentrate on communicating basic ideas of risk to the public.

9.6.2   Building a Risk Assessment Model

Building a risk assessment model involves all three steps outlined in the definition of risk assessment (hazard identification, dose-response assessment, and exposure assessment). Once these steps have been completed, some form of Equation (9.8) is usually used to estimate the risk for a particular population of concern.

The form of the two terms on the right-hand side of Equation (9.8) may be very complex. Estimation of dose involves identifying sources of exposure, postulating pathways of exposure from these sources, estimating exposure concentrations, and estimating the resulting dose for a given exposure. Estimation of dose-response involves using information from controlled laboratory experiments on animals and/or epidemiological studies. Given a set of dose-response data, there are several possible statistical models that can be used to fit these data, including tolerance distribution models, mechanistic models, linear-quadratic-exponential models, and time-to-response models (Hallenbeck, 1993, Chapter 4).

Probably the biggest controversy in risk assessment involves the extrapolation of dose-response data from high-dose to low-dose and from one species to another (e.g., between mice and humans). Rodricks (1992) discusses these problems in detail. A very recent example of this problem is the case of saccharin, which was shown in the late 1970s to produce bladder tumors in male rats that were fed extremely large concentrations of the chemical. These studies led the FDA to call for a ban on saccharin, but Congress placed a moratorium on the ban that was renewed periodically. A little over two decades later, the National Institute of Environmental Health Sciences states that new studies show "no clear association" between saccharin and human cancer (The Seattle Times, Tuesday, May 16, 2000) and has taken saccharin off of its list of cancer-causing chemicals.

Many risk assessment models have the general form

$$Risk = Dose \times \Pr(\textit{Effect per Unit Dose})$$

$$(9.9)$$

$$= h(X_1, X_2, \ldots, X_k) \times CPF$$

That is, the risk is assumed to be proportional to dose, and the dose term is a function of several input variables. For example, USEPA (1991b, p. 2) uses the following general equation for intake (here intake is equivalent to exposure and dose):

$$Intake = \frac{C \times IR \times EF \times ED}{BW \times AT} \tag{9.10}$$

where C is the chemical concentration, IR is the intake or contact rate, EF is the exposure frequency, ED is the exposure duration, BW is body weight, and AT is the averaging time (equal to exposure duration for non-carcinogens and 70 years for carcinogens).

Usually, many of the input variables and sometimes the cancer potency factor (CPF) in Equation (9.9) are themselves assumed to be random variables because they exhibit inherent heterogeneity (variability) within the population (e.g., body weight, fluid intake, etc.), and because there is a certain amount of uncertainty associated with their values. The choice of what distribution to use for each of the input variables is based on a combination of available data and expert judgment.

9.6.3 Example: Cancer Risk from PCE in Groundwater

McKone and Bogen (1991) present a risk assessment model for lifetime cancer risk from exposure to perchloroethylene (PCE) in groundwater in California. In order to estimate dose, they estimated the concentration of PCE in California water supplies, identified three pathways of exposure (ingestion of tap water, indoor inhalation from tap water, and dermal contact from bathing and showering), estimated concentrations and absorption rates for each pathway, and estimated actual dose based on exposure concentration for each pathway. To estimate the cancer potency factor (CPF), McKone and Bogen (1991) used eight rodent-bioassay data sets and a multistage dose-response extrapolation model. They assumed the input variables for dose and the CPF follow certain probability distributions, and they present two point estimates of risk (using the mean value of each input variable and the 95th percentile of each input variable) as well as a distribution of risk based on a Monte Carlo simulation.

9.6.4 Example: Incremental Lifetime Cancer Risk from Ingesting Soil Contaminated with Benzene

Thompson et al. (1992) present a simplified case study of a risk assessment for incremental lifetime cancer risk (ILCR) for children (ages 8 to 18) ingesting soil contaminated with benzene. They use the following model:

$$ILCR = Dose \times CPF$$

$$(9.11)$$

$$= \frac{Cs \cdot SIngR \cdot RBA \cdot DpW \cdot WpY \cdot YpL \cdot CF}{BW \cdot DinY \cdot YinL} \times CPF$$

Symbol	Variable Name	Units	Distribution	Point Estimate
Cs	Benzene Concentration	mg/kg	Lognormal (0.84, 0.77)	3.39
$SingR$	Soil Ingestion Rate	mg/dy	Lognormal (3.44, 0.80)	50
RBA	Relative Bioavailability			1
DpW	Exposure Days per Week	dy/wk		1
WpY	Exposure Weeks per Year	wk/yr		20
YpL	Exposure Years per Life	yr/life		10
CF	Conversion Factor	kg/mg		1×10^{-6}
BW	Average Body Weight	kg	Normal (47, 8.3)	47
$DinY$	Days in a Year	dy/yr		364
$YinL$	Years in a Lifetime	yr/life		70
CPF	Cancer Potency Factor	(kg dy)/mg	Lognormal (-4.33, 0.67)	0.029

Table 9.2. Description of terms in Equation (9.11) (Thompson et al., 1992).

Table 9.2 explains what the input variables are and their assumed distributions. In this table, the normal distribution is parameterized by its mean and standard deviation, and the lognormal distributions are parameterized by the mean and standard deviation of the log-transformed random variable. The point estimates for the four input variables with associated probability distributions, benzene concentration (Cs), soil ingestion rate ($SIngR$), average body weight (BW), and cancer potency factor (CPF) are, respectively, the 95% upper confidence limit for the mean, the 72nd percentile, the mean, and the 88th percentile. Thompson et al. (1992) used the model in Equation (9.11) (along with a model for ILCR due to dermal contact with the soil) to demonstrate how to use Monte Carlo simulation to perform risk assessment.

9.6.5   Point Estimates of Risk

USEPA (1989c, pp. 6-4 to 6-5) states that

> Actions at Superfund sites should be based on an estimate of *the reasonable maximum exposure (RME)* expected to occur under both *current* and *future* land-use conditions. The reasonable maximum exposure is defined here as the highest exposure that is reasonably expected to occur at a site. RMEs are estimated for individual pathways. If a population is exposed via more than one pathway, the combination of exposure pathways also must represent an RME. ... The intent of the RME is to estimate a conservative exposure case (i.e., well above the average case) that is still within the range of possible exposures.

In the context of estimating the exposure term of a risk assessment model, the document further specifies that

> Each intake variable in the equation has a range of values. For Superfund exposure assessments, intake variable values for a given pathway should be selected so that the combination of all intake variables results in an estimate of the reasonable maximum exposure for that pathway. ... Under this approach, some intake variables may not be at their individual maximum values but when in combination with other variables will result in estimates of the RME.

(USEPA, 1989c, p. 6-19). The document then goes on to suggest using the 95% upper confidence limit of the mean for exposure concentration, the 95th or 90th percentile for contact rate, the 95th percentile for exposure frequency and duration, and the average body weight over the exposure period (USEPA, 1989c, pp. 6-19, 6-22). This guidance is intended to yield an estimate of risk that is both "protective and reasonable." USEPA (1991b, p. 2) reiterates these suggestions.

The above type risk estimate is a "conservative" ***point estimate of risk***. This type of estimate of risk has been frequently criticized for the following reasons:

- There is no way of assessing the degree of conservatism in this kind of estimate (Thompson et al., 1992).
- The input variables may be set to a specific combination that will rarely, if ever, occur in reality (Thompson et al., 1992).
- Traditional sensitivity analyses (one-at-a-time-deviations from the baseline case) are meaningless for this baseline case since many of the input variables are already at or close to their maximum values (Thompson et al., 1992).
- This estimate of risk is often larger by one or several orders of magnitude than the 95th percentile of the population risk (McKone and Bogen, 1991; Burmaster and Harris, 1993; Cullen, 1994).

9.6.6   Probabilistic Risk Assessment

Because of the shortcomings of point estimates of risk, most practitioners in the field of risk assessment now advocate presenting not only the results of points estimates but also the whole distribution of risk and the characteristics of this distribution (Eschenroeder and Faeder, 1988; McKone and Bogen, 1991; Thompson et al., 1992; Burmaster and Harris, 1993; Finley and Paustenbach, 1994; Smith, 1994; McKone, 1994). Presenting the distribution of the output variable (risk in this case) and characterizing this distribution is in fact the definition of uncertainty analysis (Iman and Helton, 1988). *Probabilistic risk assessment is simply uncertainty analysis applied to a risk assessment model that assumes some of the input variables in the model are random variables.*

Sometimes the distribution of risk can be derived analytically for simple models (Springer, 1979; Slob, 1994). Usually, however, the distribution of risk must be derived by Monte Carlo simulation. We will give a specific example later in this chapter.

9.6.7 Guidelines for Conducting and Reporting a Probabilistic Risk Assessment

Because of the complexity of probabilistic risk assessment, it is important to carefully conduct the risk assessment, document exactly how the risk assessment was performed, and effectively communicate the results. Hoaglin and Andrews (1975) suggest certain procedures to follow when reporting the results of a Monte Carlo simulation. More recent references specific to probabilistic risk assessment include Burmaster and Anderson (1994), Vose (1996, Chapter 12), and USEPA (1997). The following guidelines are based on these last three references.

Conducting a Probabilistic Risk Assessment

1. Identify the purpose and scope of the risk assessment, and define the endpoints.
2. Create a model or several models for the risk assessment based on the three steps outlined in the definition of risk assessment (hazard identification, dose-response assessment, and exposure assessment). Document the assumptions of each model.
3. Conduct preliminary sensitivity analyses to determine if structural differences (different models) have substantial effects on the output distribution (in both the region of central tendency and the tails). For a given model, determine which input variables contribute the most to the variability of the output distribution. If warranted, determine the effects of changing the shape of the input distributions and/or accounting for correlations between the input variables (Smith et al., 1992; Bukowski et al., 1995; Hammed and Bedient, 1997).
4. Simplify the Monte Carlo simulation by setting unimportant input variables (determined from the previous step) to fixed values.
5. Develop distributions for important input variables based on available data and expert judgment. Document the sources of the data sets and what assumptions were used to derive the input distributions. Perform goodness-of-fit tests and graphical assessments of goodness-of-fit.

6. Quantify and distinguish between the variability and uncertainty in the input parameters. For important input parameters, it may be necessary to gather additional data to reduce uncertainty.

7. Perform the Monte Carlo simulations, always recording the random number seeds. Perform sensitivity analyses to determine which variables or group of variables contribute substantially to the variability of the output variable (e.g., exposure or risk). Distinguish between variability and uncertainty in the input parameters (one way to do this is to model the input distributions as mixture distributions; see Hoffman and Hammonds, 1994 and Burmaster and Wilson, 1996).

8. Verify the numerical stability of the Monte Carlo simulation results by performing at least two simulations (using different starting random seeds, of course) or dividing the simulation into non-overlapping subsequences and comparing results.

Presenting the Results of a Probabilistic Risk Assessment

1. Taylor the document to the client. Results presented to a risk manager will include many technical details that are usually omitted for presentations to the general public. Results presented to a risk manager should be detailed enough so that the risk assessment may be independently duplicated and verified.

2. Use a tiered presentation style. Put the question and results up front, put details in appendices.

3. Use both a histogram and empirical cumulative distribution function plot to present the distribution of risk on one page, using identical horizontal scales for each plot. On each plot, identify the mean, the 95th percentile, and the point estimate of risk based on established regulatory guidance. Accompany these plots with a table of summary statistics (e.g., minimum, 5th percentile, median, mean, 95th percentile, and maximum).

4. Present and explain the model. Include all equations, describe all terms and their units of measure, and, where appropriate, include schematic diagrams documenting the structure of the model. Outline model assumptions and shortcomings. Document references for the model.

5. Document how each input distribution was selected, including sources of any relevant data sets. Include results of goodness-of-fit tests and graphical assessments of goodness-of-fit. Also discuss the relative contributions of uncertainty and variability for each input distribution. For each input distribution, present a plot of the probability density function and cumulative distribution function on one page, using identical horizontal axes for each plot. Indicate the location of the mean on each plot. Also include a table of relevant statistics for the input distribution, including the minimum, 5th percentile, median, mean, 95th percentile, and maximum.

6. Discuss the results of sensitivity and uncertainty analyses, including the effects of changing the shape of input distributions and/or including

correlations among the input variables, as well as analyses that separate uncertainty from variability.

7. Document the software that was used to perform the Monte Carlo simulation, and present the name and statistical quality (i.e., the period) of the random number generator used by the software.

9.7 Risk Assessment for Benzene-Contaminated Soil

In this section we will give an abbreviated example of performing a risk assessment based on Thompson et al.'s (1992) model for incremental lifetime cancer risk (ILCR) for children ingesting soil contaminated with benzene. The model is described in Equation (9.11) and Table 9.2. The information in this section is also contained in the ENVIRONMENTALSTATS for S-PLUS help file *Glossary: Risk Assessment – Using ENVIRONMENTALSTATS for S-PLUS for Probabilistic Risk Assessment.*

9.7.1 Point Estimates of Exposure and Risk

Using the point estimates of the input variables shown in Table 9.2, the point estimates of exposure is about 3×10^{-8} mg/(kg·d), and the point estimate of incremental lifetime cancer risk is about 8×10^{-10} (8 in 10 billion). The point estimate of soil ingestion rate (*SIngR*) shown in Table 9.2 is the 72nd percentile, and the point estimate of the cancer potency factor (*CPF*) shown in Table 9.2 is the 88th percentile. If instead we follow the suggestions of (USEPA, 1989c) and use the 95th percentiles for these two input variables, the point estimates of exposure and risk increase, as shown in Table 9.3.

SIngR	*CPF*	Exposure	ILCR
72nd Percentile	88th Percentile	3×10^{-8}	8×10^{-10}
95th Percentile	95th Percentile	7×10^{-8}	3×10^{-9}

Table 9.3. Point estimates of exposure and incremental lifetime cancer risk (ILCR).

Menu

To compute the point estimates of exposure and risk shown in the first row of Table 9.3, follow these steps.

1. Create a data frame called **benzene.df** with one row that contains the values of the point estimates shown in Table 9.2. The names of the columns in this data frame should be the same as the names shown in the column labeled **Symbol** in Table 9.2.

2. Find **benzene.df** in the Object Explorer.

3. On the S-PLUS menu bar, make the following menu choices: **Data>Transform**. This will bring up the Transform dialog box.
4. For Data Set select **benzene.df**, for Target Column type **exposure.pt.est.1**, in the Expression box type **(Cs * SIngR * RBA * DpW * WpY * YpL * CF) / (BW * DinY * YinL)**, and click **Apply**.
5. For Data Set select **benzene.df**, for Target Column type **ilcr.pt.est.1**, in the Expression box type **exposure.pt.est.1 * CPF**, and click **OK**.
6. Move to the **benzene.df** data window, right-click on the **exposure.pt.est.1** column, and choose **Properties**. For Format Type select **Scientific**, then click **OK**.
7. Repeat the last step for the column **ilcr.pt.est.1**.

To compute the point estimates of exposure and risk shown in the second row of Table 9.3, follow these steps.

1. Find **benzene.df** in the Object Explorer.
2. On the S-PLUS menu bar, make the following menu choices: **Data>Transform**. This will bring up the Transform dialog box.
3. For Data Set select **benzene.df**, for Target Column type **SIngR.95**, in the Expression box type **qlnorm(0.95, 3.44, 0.8)**, and click **Apply**.
4. For Data Set select **benzene.df**, for Target Column type **CPF.95**, in the Expression box type **qlnorm(0.95, -4.33, 0.67)**, and click **Apply**.
5. For Data Set select **benzene.df**, for Target Column type **exposure.pt.est.2**, in the Expression box type **(Cs * SIngR.95 * RBA * DpW * WpY * YpL * CF) / (BW * DinY * YinL)**, and click **Apply**.
6. For Data Set select **benzene.df**, for Target Column type **ilcr.pt.est.2**, in the Expression box type **exposure.pt.est.2 * CPF.95**, and click **OK**.
7. Move to the **benzene.df** data window, right-click on the **exposure.pt.est.2** column, and choose **Properties**. For Format Type select **Scientific**, then click **OK**.
8. Repeat the last step for the column **ilcr.pt.est.2**.

Command

To compute the point estimates of exposure and risk, type the following commands. First we will create a function called `exposure.fcn` to model exposure, and then we will create a function called `ilcr.fcn` to model incremental lifetime cancer risk. Finally, we will plug in the points estimates of the input variables.

```
> exposure.fcn <- function(Cs, SIngR, BW, RBA=1, DpW=1,
    WpY=20, YpL=10, CF=1e-006, DinY=364, YinL=70) {
      (Cs * SIngR * RBA * DpW * WpY * YpL * CF) /
      (BW * DinY * YinL)
    }
```

```
> ilcr.fcn <- function(Cs, SIngR, BW, CPF, RBA=1, DpW=1,
    WpY=20, YpL=10, CF=1e-006, DinY=364, YinL=70) {
      exposure.fcn(Cs=Cs, SIngR=SIngR, BW=BW,
        RBA=RBA, DpW=DpW, WpY=WpY, YpL=YpL, CF=CF,
        DinY=DinY, YinL=YinL) * CPF
  }
> exposure.pt.est.1 <- exposure.fcn(Cs=3.39, SIngR=50,
    BW=47)
> ilcr.pt.est.1 <- ilcr.fcn(Cs=3.39, SIngR=50, BW=47,
    CPF=0.029)
```

To compute the points estimates of exposure and risk using the 95th percentiles of soil ingestion rate and cancer potency factor, type these commands:

```
> exposure.pt.est.2 <- exposure.fcn(Cs=3.39,
    SIngR=qlnorm(0.95, 3.44, 0.8), BW=47)
> ilcr.pt.est.2 <- ilcr.fcn(Cs=3.39,
    SIngR=qlnorm(0.95, 3.44, 0.8), BW=47,
    CPF=qlnorm(0.95, -4.33, 0.67))
```

9.7.2   Characterizing the Distributions of the Input Variables

Figure 9.10 and Figure 9.11 display the hypothesized probability density function and cumulative distribution function for the benzene concentration in the soil (*Cs*). In both of these figures we have added a vertical line to indicate the value of the mean. (Note that we must use Equation (4.2) on page 86 to compute the mean of the distribution since it has been parameterized in terms of the mean and standard deviation of the log-transformed random variable.) The minimum, 5th percentile, median, 95th percentile, and maximum of the distribution are: 0, 0.7, 2.3, 8.2, and ∞. You can create pdf and cdf plots and compute population quantiles for the other three input variables that have associated distributions as well (i.e., for *SIngR*, *BW*, and *CPF*).

Menu

To create the pdf plot shown in Figure 9.10, follow these steps.

1. On the S-PLUS menu bar, make the following menu choices: **EnvironmentalStats>Probability Distributions and Random Numbers>Plot Distribution**. This will bring up the Plot Distribution Function dialog box.
2. For Distribution choose **Lognormal**. For meanlog type **0.84** and for sdlog type **0.77**. For Right Tail Cutoff type **0.005**. For Plot choose **PDF**. **Uncheck** the Add Title box. Click **OK**.
3. Right-click on the curve and choose **Convert to Objects**.
4. Add a title and modify the *x*- and *y*-axis labels if you wish.

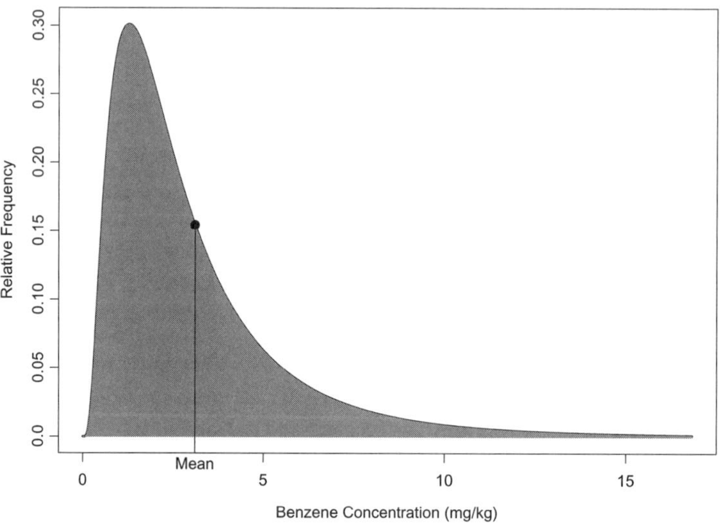

Figure 9.10. Assumed pdf for benzene concentration in soil.

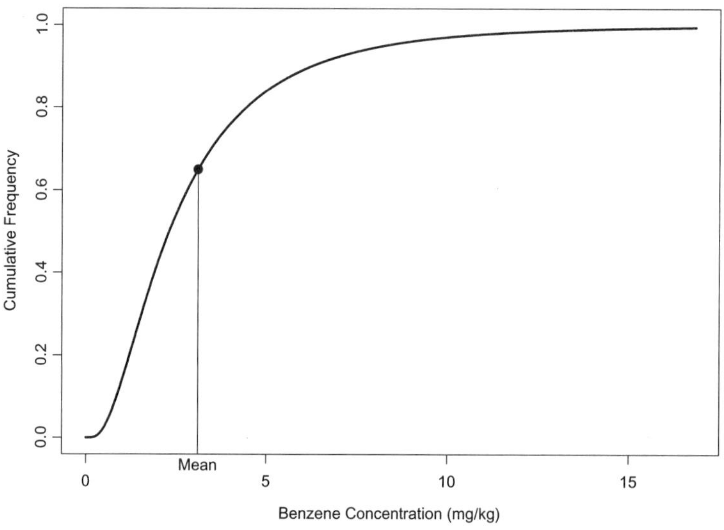

Figure 9.11. Assumed cdf for benzene concentration in soil.

5. The mean of the distribution is 3.12 mg/kg. To add a vertical line at the mean, you have to *hand place it* with the Vertical Ref Line Tool. To do this, click on the **Annotations** button to bring up the Annotations palette, then click on the **Vertical Ref Line Tool**. Move the cursor to the bottom of the *x*-axis and click where you think 3.12 is. This produces a vertical line where you clicked.

6. On the Annotations palette, click on the **Comment** tool, then click once below the vertical line you just created. **Close** the Annotations palette, then update the text box to say **Mean**.

To create the cdf plot shown in Figure 9.11, follow the same steps as above, except in Step 2 for Plot choose **CDF**.

Command

To create the pdf plot shown in Figure 9.10, type these commands.

```
> pdfplot(dist="lnorm",
    param.list=list(meanlog=0.84, sdlog=0.77),
    right.tail.cutoff=0.005,
    xlab="Benzene Concentration (mg/kg)",
    main="Assumed PDF for Benzene Concentration")
> avg <- exp(0.84 + (0.77^2)/2)
> f.avg <- dlnorm(avg, 0.84, 0.77)
> segments(avg, par("usr")[3], avg, f.avg)
> points(avg, f.avg, pch=16, cex=1.25*par("cex"))
> mtext("Mean", side=1, at=avg)
```

To create the cdf plot shown in Figure 9.11, type these commands.

```
> cdfplot(dist="lnorm",
    param.list=list(meanlog=0.84, sdlog=0.77),
    right.tail.cutoff=0.005,
    xlab="Benzene Concentration (mg/kg)",
    main="Assumed CDF for Benzene Concentration")
> F.avg <- plnorm(avg, 0.84, 0.77)
> segments(avg, par("usr")[3], avg, F.avg)
> points(avg, F.avg, pch=16, cex=1.25*par("cex"))
> mtext("Mean", at=avg, side=1)
```

9.7.3 Simulating the Input Variables

In this section we will create a matrix or data frame to hold simulated observations of the input variables that have associated probability distributions. We will generate 10,000 observations for each of the variables benzene concentration (*Cs*), soil ingestion rate (*SIngR*), body weight (*BW*), and cancer potency factor (*CPF*) according to their specified distributions as shown in Table 9.2, assuming these variables are independent. In reality, there is probably a relationship between soil ingestion rate and body weight so these two variables are not independent, and we could instead simulate these four variables so that there is some specified rank correlation between *SIngR* and *BW*. Both Smith et al. (1992) and Bukowski et al. (1995) show that for many circumstances, however, the shapes of the chosen input probability distributions have much more effect on the resultant distribution of risk than does including or excluding correlations between input variables.

Here we will simulate the observations based on simple random sampling (SRS). We could instead use Latin Hypercube sampling (LHS), but as we stated earlier in this chapter, in cases where it is fairly easy to generate tens of thousands of Monte Carlo trials, Latin Hypercube sampling may not offer any real advantage. In fact, in this particular example there is very little difference in outcome in terms of standard summary statistics between SRS and LHS.

Menu

To generate the $10,000 \times 4$ data frame **simulate.df** of simulated observations based on simple random sampling, follow these steps.

1. On the S-PLUS menu bar, make the following menu choices: **EnvironmentalStats>Probability Distributions and Random Numbers>Random Numbers>Multivariate>Based on Rank Correlation**. This will bring up the Multivariate Random Number Generation dialog box.
2. For Sample Size type **10000**, for Set Seed with type **47**, for Number of Vars select **4**, **uncheck** the Print Results box, and for Save As type **simulate.df**.
3. Click on the **Variable 1** tab. For Variable Name type **Cs**, for Distribution Type select **Parametric**, for Sampling Method select **Random**, for Distribution select **Lognormal**, for meanlog type **0.84**, and for sdlog type **0.77**.
4. Click on the **Variable 2** tab. For Variable Name type **SIngR**, for Distribution Type select **Parametric**, for Sampling Method select **Random**, for Distribution select **Lognormal**, for meanlog type **3.44**, and for sdlog type **0.8**.
5. Click on the **Variable 3** tab. For Variable Name type **BW**, for Distribution Type select **Parametric**, for Sampling Method select **Random**, for Distribution select **Normal**, for mean type **47**, and for sd type **8.3**.
6. Click on the **Variable 4** tab. For Variable Name type **CPF**, for Distribution Type select **Parametric**, for Sampling Method select **Random**, for Distribution select **Lognormal**, for meanlog type **-4.33**, and for sdlog type **0.67**.
7. Click **OK**.

Command

To generate the 10,000 × 4 matrix `input.mat` of simulated observations based on simple random sampling, type this command.

```
> input.mat <- simulate.mv.matrix(10000,
     distributions=c(Cs    = "lnorm",
                     SIngR = "lnorm",
                     BW    = "norm",
                     CPF   = "lnorm"),
  param.list = list(
    Cs      = list(meanlog =  0.84, sdlog = 0.77),
    SIngR   = list(meanlog =  3.44, sdlog = 0.80),
    BW      = list(mean    = 47,    sd    = 8.3),
    CPF     = list(meanlog = -4.33, sdlog = 0.67)),
  cor.mat = diag(4), seed = 47)
```

9.7.4 Characterizing the Simulated Distributions of the Input Variables

It is a good idea to look at the actual distributions of the input variables that were generated with the Monte Carlo simulation, and also at the relationships between the variables. Based on these analyses, you may decide to omit certain cases before computing the simulated distributions of exposure and risk. Here are some basic summary statistics for the simulated data:

```
> round(apply(input.mat, 2, summary), 2)
             Cs   SIngR     BW   CPF
    Min.   0.08    0.93  19.07  0.00
 1st Qu.   1.37   18.14  41.44  0.01
  Median   2.35   30.74  47.08  0.01
    Mean   3.14   42.44  47.07  0.02
 3rd Qu.   3.96   52.61  52.66  0.02
    Max.  34.35  773.60  77.70  0.18
```

Based on these summary statistics, we may or may not want to exclude certain cases. For example, the model in Equation (9.11) applies to children ages 8–18. If an average body weight (*BW*) of less than 20 kg (about 44 pounds) does not make sense for this scenario, then we should exclude all cases where the value of *BW* is less than 20. (This is essentially specifying a truncated normal distribution for average body weight.)

Figure 9.12 and Figure 9.13 display the empirical pdf and cdf of benzene concentration (*Cs*). Compare these figures to Figure 9.10 and Figure 9.11. Figure 9.14 displays the pair-wise scatterplots of all four simulated variables (see the S-PLUS documentation for information on how to create pair-wise scatterplots).

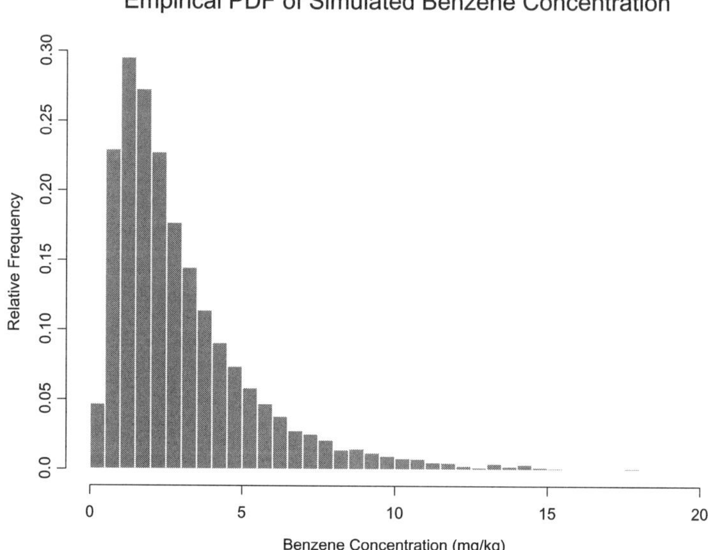

Figure 9.12. Probability histogram of simulated values for benzene concentration in soil.

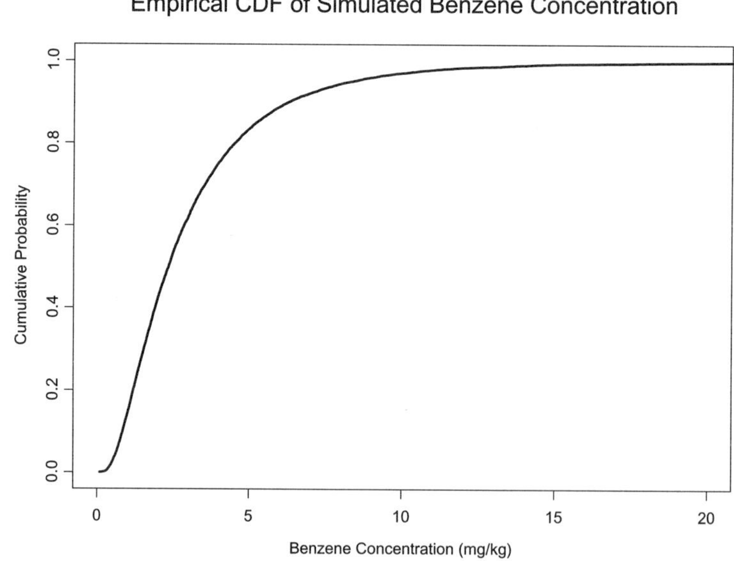

Figure 9.13. Empirical cdf of simulated values for benzene concentration in soil.

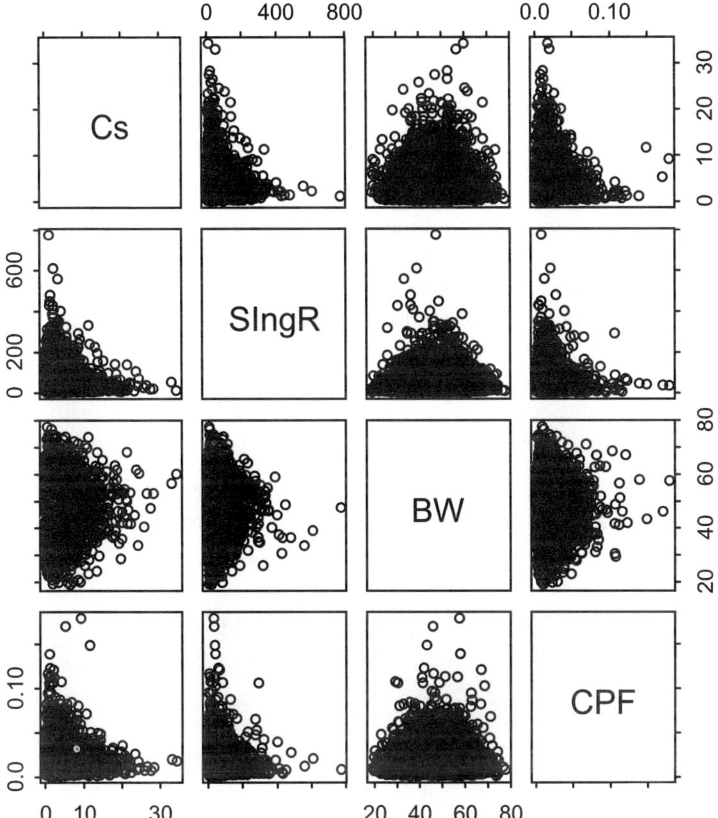

Figure 9.14. Pair-wise scatterplots for the four input variables with assumed probability distributions.

9.7.5   Characterizing the Simulated Distributions of Exposure and Risk

Once you are satisfied with the input distributions you have chosen and the simulated values, you can compute the exposure and risk for each outcome of the simulated variables. Then you need to characterize the distribution of exposure or risk.

To compute the simulated exposure and risk using the S-PLUS menu, you make the selection **Data>Transform**, as we will show below. To compute the simulated exposure and risk using the Command or Script Window, you can use the ENVIRONMENTALSTATS for S-PLUS function `eval.expr` (evaluates an

S-PLUS expression) to compute the empirical distribution of risk or exposure. The first argument to `eval.expr` is any legitimate S-PLUS expression. This expression should be the formula for exposure or risk, or an expression that calls a function that computes the exposure or risk. The second argument to `eval.expr` is a matrix, data frame, or list that contains simulated random numbers. Each column of the matrix or data frame or each component of the list should have a name corresponding to one of the variables that are used in the expression for risk or exposure.

Figure 9.15 displays the empirical probability distribution function (i.e., the histogram) of incremental lifetime cancer risk (ILCR) based on the Monte Carlo simulation. Figure 9.16 displays the empirical cumulative distribution function. In both of these plots we have identified the mean and 95th percentile of the simulated distribution, along with the point estimate of risk shown in the first row of Table 9.3. The average ILCR is about 4 in 10 billion, the point estimate is about 8 in 10 billion, and the 95th percentile is about 1 in one billion.

Here are some summary statistics for the simulated ILCR distribution, where we have multiplied by 1×10^9 to make it easier to read.

```
> round(summary(1e9 * simulated.ilcr), 3)
  Min. 1st Qu. Median Mean 3rd Qu.  Max.
 0.001   0.066  0.157 0.37   0.386 15.93


> round(quantile(1e9 * simulated.ilcr,
    c(0, 0.05, 0.5, 0.95, 1)), 3)
    0%    5%    50%    95%    100%
 0.001 0.018 0.157 1.364 15.934
```

Thompson et al.'s point estimate of risk shown in the first row of Table 9.3 falls at about the 90th percentile of the simulated ILCR, whereas the more conservative point estimate of risk shown in the second row of Table 9.3 falls at about the 98th percentile of the simulated ILCR:

```
> round(100 * pemp(c(ilcr.pt.est.1, ilcr.pt.est.2),
    obs=simulated.ilcr), 0)
[1] 90 98
```

Menu

To compute the simulated distribution for exposure and ILCR, follow these steps.

1. Find **simulate.df** in the Object Explorer.
2. On the S-PLUS menu bar, make the following menu choices: **Data>Transform**. This will bring up the Transform dialog box.
3. For Data Set select **simulate.df**, for Target Column type **exposure**, in the Expression box type **(Cs * SIngR * 1 * 1 * 20 * 10 * 1e-6) / (BW * 364 * 70)**, and click **Apply**.

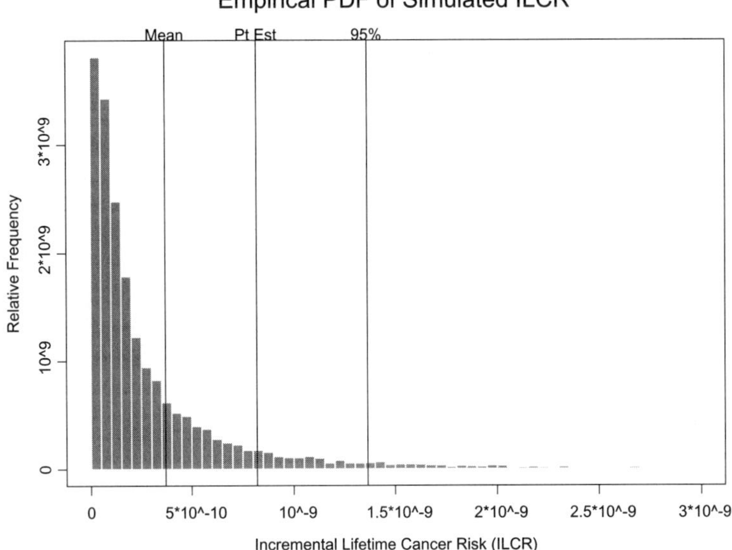

Figure 9.15. Empirical probability density of ILCR.

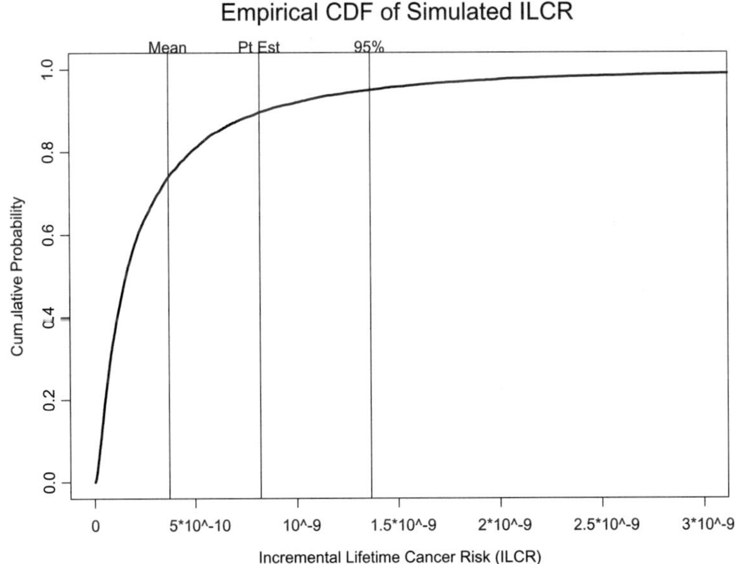

Figure 9.16. Empirical cdf of ILCR.

4. For Data Set select **simulate.df**, for Target Column type **ILCR**, in the Expression box type **exposure * CPF**, and click **OK**.
5. Move to the **simulate.df** data window, right-click on the **exposure** column, and choose **Properties**. For Format Type select **Scientific**, then click **OK**.
6. Repeat the last step for the **ILCR** column.

To computed the mean and 95th percentile of the simulated distribution of ILCR, follow these steps.

1. On the S-PLUS menu bar, make the following menu choices: **Data>Transform**. This will bring up the Transform dialog box.
2. For Data Set select **simulate.df**, for Target Column type **ILCR.avg**, in the Expression box type **mean(ILCR)**, and click **Apply**.
3. For Data Set select **simulate.df**, for Target Column type **ILCR.95.pct**, in the Expression box type **quantile(ILCR, 0.95)**, and click **OK**.
4. Move to the **simulate.df** data window, right-click on the **ILCR.avg** column, and choose **Properties**. For Format Type select **Scientific**, then click **OK**.
5. Repeat the last step for the **ILCR.95.pct** column.

To plot the histogram of the simulated distribution of ILCR and add the mean, point estimate, and 95th percentile, follow these steps.

1. On the S-PLUS menu bar, make the following menu choices: **Graph>2D Plot**. This will bring up the Insert Graph dialog box. Under Axes Type column, **Linear** should be highlighted. Under the Plot Type column select **Histogram** and click **OK**. (Alternatively, left-click on the **2D** Plots button, then left-click on the **Histogram** button.)
2. The Histogram/Density dialog box should appear. For Data Set select **simulate.df**. For x Columns select **ILCR**.
3. Click on the **Options** tab. For Output Type select **Density**. For Lower Bound type **0**, and for Upper Bound type **3.5e-9**. For Number of Bars type **75**. Click **OK**.
4. Add a title and modify the x- and y-axis labels if you wish. Modify the x- and y-axis tick marks and labels if you wish.
5. The mean of the distribution is 3.7×10^{-10}. To add a vertical line at the mean, you have to *hand place it* with the Vertical Ref Line Tool. To do this, click on the **Annotations** button to bring up the Annotations palette, then click on the **Vertical Ref Line Tool**. Move the cursor to the bottom of the x-axis and click where you think 3.7×10^{-10} is. This produces a vertical line where you clicked.
6. On the Annotations palette, click on the **Comment** tool, then click once at the top of the vertical line you just created. **Close** the Annotations palette, then update the text box to say **Mean**.
7. Repeat Steps 5 and 6 to place a line at the point estimate and 95th percentile.

Command

To compute the simulated distribution for exposure, type one of the following two commands.

```
> simulated.exposure <- eval.expr(
    (Cs * SIngR * 1 * 1 * 20 * 10 * 1e-6) /
    (BW * 364 * 70), data=input.mat)
> simulated.exposure <- eval.expr(
    exposure.fcn(Cs, SIngR, BW), data=input.mat)
```

To compute the simulated distribution for ILCR, type one of the following two commands.

```
> simulated.ilcr <- eval.expr(
    ((Cs * SIngR * 1 * 1 * 20 * 10 * 1e-6) /
    (BW * 364 * 70)) * CPF, data=input.mat)
> simulated.ilcr <- eval.expr(
    ilcr.fcn(Cs, SIngR, BW, CPF), data=input.mat)
```

To computed the mean and 95th percentile of the simulated distribution of ILCR, type these commands.

```
> ilcr.avg <- mean(simulated.ilcr)
> ilcr.95.pct <- quantile(simulated.ilcr, 0.95)
```

To plot the histogram of the simulated distribution of ILCR and add the mean, point estimate, and 95th percentile, type these commands.

```
> hist(simulated.ilcr, nclass=500, prob=T,
    xlim=c(0, 3e-9),
    xlab="Incremental Lifetime Cancer Risk (ILCR)",
    ylab="Relative Frequency",
    main="Empirical PDF of Simulated ILCR")
> abline(v=ilcr.avg)
> abline(v=ilcr.95.pct)
> abline(v=ilcr.pt.est.1)
> box()
> mtext("Mean", at=ilcr.avg)
> mtext("95%", at=ilcr.95.pct)
> mtext("Pt Est", at=ilcr.pt.est.1)
```

To plot the empirical cdf of the simulated distribution of ILCR and add the mean, point estimate, and 95th percentile, type these commands.

```
> ecdfplot(simulated.ilcr, xlim = c(0, 3e-9),
    xlab="Incremental Lifetime Cancer Risk (ILCR)",
    main="Empirical CDF of Simulated ILCR")
> abline(v=ilcr.avg)
> abline(v=ilcr.95.pct)
```

```
> abline(v=ilcr.pt.est.1)
> mtext("Mean", at=ilcr.avg)
> mtext("95%", at=ilcr.95.pct)
> mtext("Pt Est", at=ilcr.pt.est.1)
```

9.7.6 Sensitivity Analysis

As discussed earlier in this chapter, there are several methods of performing sensitivity analysis for risk assessment. For this example here, we will simply look at the change in distribution of incremental lifetime cancer risk as we let one input variable vary and keep the other variables fixed at their point estimates (shown in Table 9.2). Figure 9.17 displays the original histogram for ILCR, along with the four others produced by varying only average body weight (BW), benzene concentration in soil (Cs), soil ingestion rate ($SIngR$), or cancer potency factor (CPF) one at a time. Similar plots appear in Thompson et al. (1992).

Based on these figures, you can see that average body weight (BW) has the smallest effect on the variability of ILCR compared to the other three input variables. Also, the bulk of the distribution based on varying any of the other three variables is below the point estimate of risk, with the cancer potency factor (CPF) having the greatest influence.

Menu

To produce the simulated distribution of ILCR where only average body weight (BW) is varied, follow these steps.

1. On the S-PLUS menu bar, make the following menu choices: **Data>Transform**. This will bring up the Transform dialog box.
2. For Data Set select **simulate.df**, for Target Column type **ILCR.Vary.BW**, in the Expression box type **0.029 * (3.39 * 50 * 1 * 1 * 20 * 10 * 1e-6) / (BW * 364 * 70)**, and click **OK**.
3. Move to the **simulate.df** data window, right-click on the **ILCR.Vary.BW** column, and choose **Properties**. For Format Type select **Scientific**, then click **OK**.
4. On the S-PLUS menu bar, make the following menu choices: **Graph>2D Plot**. This will bring up the Insert Graph dialog box. Under Axes Type column, **Linear** should be highlighted. Under the Plot Type column select **Histogram** and click **OK**. (Alternatively, left-click on the **2D** Plots button, then left-click on the **Histogram** button.)
5. The Histogram/Density dialog box should appear. For Data Set select **simulate.df**. For x Columns select **ILCR.Vary.BW**.
6. Click on the **Options** tab. For Output Type select **Density**. For Lower Bound type **0**, and for Upper Bound type **3.5e-9**. For Number of Bars type **75**. Click **OK**.

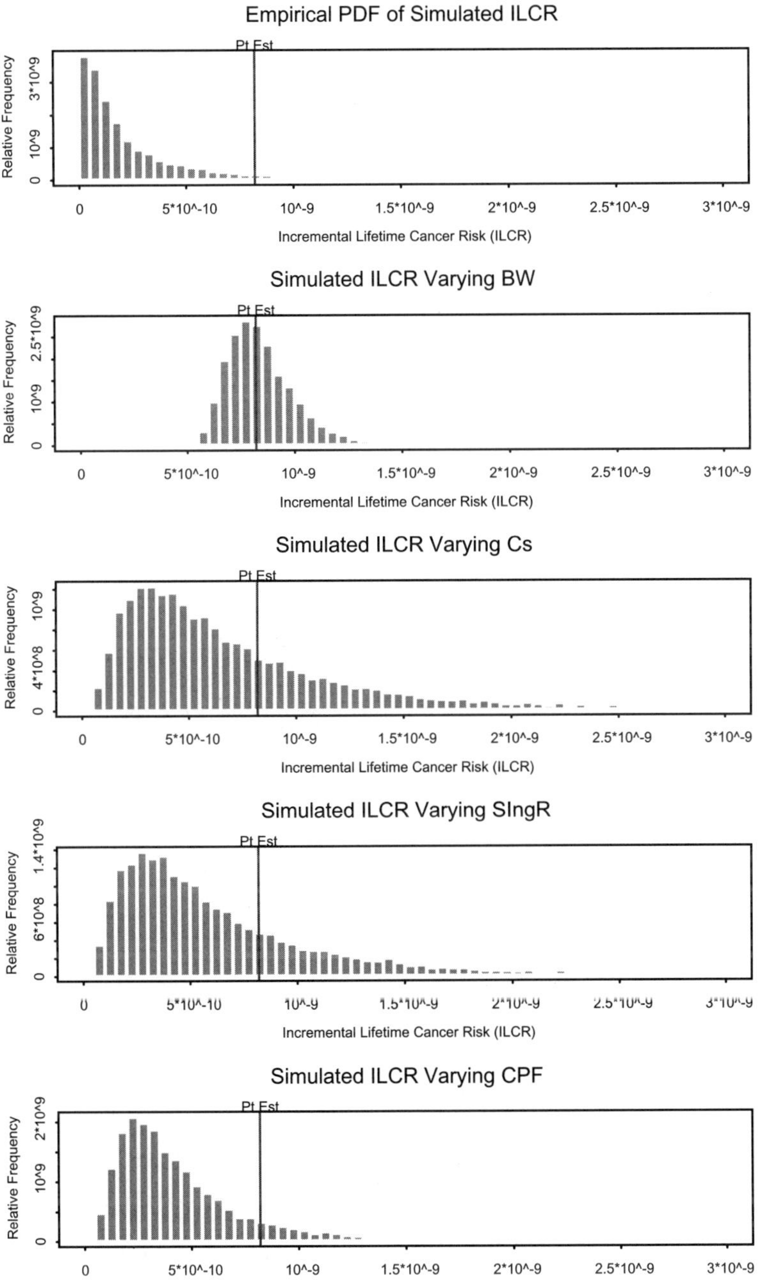

Figure 9.17. Sensitivity analysis for the model for ILCR.

7. Add a title and modify the *x*- and *y*-axis labels if you wish. Modify the *x*- and *y*-axis tick marks and labels if you wish.

8. The point estimate is 8.2×10^{-10}. To add a vertical line at the point estimate, you have to *hand place it* with the Vertical Ref Line Tool. To do this, click on the **Annotations** button to bring up the Annotations palette, then click on the **Vertical Ref Line Tool**. Move the cursor to the bottom of the *x*-axis and click where you think 8.2×10^{-10} is. This produces a vertical line where you clicked.

9. On the Annotations palette, click on the **Comment** tool, then click once at the top of the vertical line you just created. **Close** the Annotations palette, then update the text box to say **Pt Est**.

To produce the other histograms, follow the same steps as above but in Step 2 call the column the appropriate name and use the relevant point estimates for the fixed variables. For example, to produce the histogram where benzene concentration (*Cs*) is varied, in Step 2 for Target Column type **ILCR.Vary.Cs**, and in the Expression box type **0.029 * (Cs * 50 * 1 * 1 * 20 * 10 * 1e-6) / (47 * 364 * 70)**.

Command

To produce the simulated distribution of ILCR where only average body weight (*BW*) is varied, type these commands.

```
> hist(ilcr.fcn(Cs=3.39, SIngR=50,
    BW=input.mat[, "BW"], CPF=0.029), nclass=100, prob=T,
    xlim=c(0, 3e-9),
    xlab="Incremental Lifetime Cancer Risk (ILCR)",
    ylab="Relative Frequency",
    main="Simulated ILCR Varying BW")
> abline(v = ilcr.pt.est.1)
> box()
> mtext("Pt Est", at=ilcr.pt.est.1)
```

To produce the other histograms, type the same commands but subscript the appropriate column from `input.mat`.

9.7.7   Uncertainty Analysis

As discussed earlier in this chapter, there are several methods of performing uncertainty analysis for risk assessment. We have already presented the simulated distribution of incremental lifetime cancer risk in Figures 9.15 and 9.16. The two-sided nonparametric 95% confidence interval for the median ILCR is $[1.5, 1.6] \times 10^{-10}$, and for the 95th percentile it is $[1.3, 1.4] \times 10^{-9}$. Figure 9.18 compares the empirical cumulative distribution of ILCR based on this simulation with a second simulation. As you can see, the two simulated distributions are virtually identical.

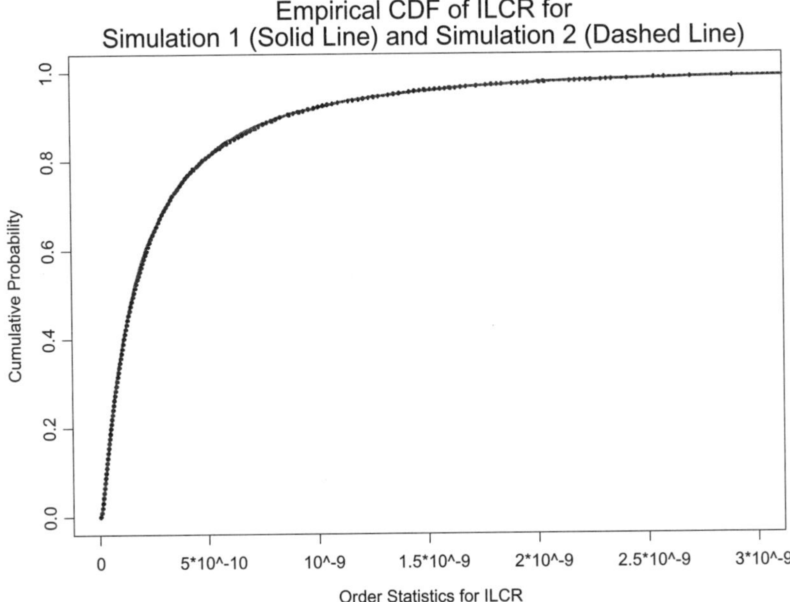

Figure 9.18. Comparison of the empirical cdf of ILCR based on two separate simulations with 10,000 Monte Carlo trials.

We could also run another simulation with an increased number of Monte Carlo trials, and/or use mixture distributions for the input variables and compare the results with our original simulation. Both of these methods of uncertainty analysis are illustrated in the ENVIRONMENTALSTATS for S-PLUS help file *Glossary: Risk Assessment – Using ENVIRONMENTALSTATS for S-PLUS for Probabilistic Risk Assessment.*

9.8 Summary

- Human and ecological risk assessment involves characterizing the exposure to a toxicant for one or several populations, quantifying the relationship between exposure (dose) and health or ecological effects (response), determining the risk (probability) of a health or ecological effect given the observed level(s) of exposure, and characterizing the uncertainty associated with the estimated risk.
- In the past, estimates of risk were often based solely on setting values of the input variables (e.g., body weight, dose, etc.) to particular point estimates and producing a single point estimate of risk, with little, if any, quantification of the uncertainty associated with the estimated risk. More recently, several practitioners have advocated "probabilistic" risk

assessment, in which the input variables are considered random variables, so the result of the risk assessment is a probability distribution for predicted risk or exposure.

- You can use S-PLUS and ENVIRONMENTALSTATS for S-PLUS to perform Monte Carlo simulation and probabilistic risk assessment. ENVIRONMENTALSTATS for S-PLUS includes menu items and functions for both simple random sampling (SRS) and Latin Hypercube sampling (LHS), as well as for generating random vectors from arbitrary distributions with a specified rank correlation matrix.

- Most risk assessment models follow the form of Equations (9.1) and (9.9), where the output variable (risk or exposure) is a function of several input variables, and some or all of the input variables are considered to be random variables.

- *Uncertainty analysis* involves describing the variability or distribution of values of the output variable that is due to the collective variation in the input variables. This description usually involves graphical displays such as histograms, empirical density plots, and empirical cdf plots, as well as summary statistics such as the mean, median, standard deviation, coefficient of variation, 95th percentile, etc.

- *Sensitivity analysis* involves determining how the distribution of the output variable changes with changes in the individual input variables. It is used to identify which input variables contribute the most to the variation or uncertainty in the output variable. Sensitivity analysis is also used in a broader sense to determine how changing the distributions of the input variables and/or their assumed correlations or even changing the form of the model affects the output.

- *Variability* refers to the inherent heterogeneity of a particular variable (parameter). For example, there is natural variation in body weight and height between individuals in a given population. *Uncertainty* refers to a lack of knowledge about specific parameters, models, or factors. We can usually reduce uncertainty through further measurement or study. We cannot reduce variability, since it is inherent in the variable.

- Sensitivity analysis methods include one-at-a-time deviations from a baseline case, factorial design and response surface modeling, and Monte Carlo simulation.

- Uncertainty analysis methods include describing the empirical distribution of risk, repeating the simulation using a different set of random numbers, and using mixture distributions.

- An important point to remember is that no matter how complex the mathematical model in Equation (9.1) or (9.9), or how extensive the uncertainty and sensitivity analyses, there is always the question of how well the mathematical model reflects reality. The only way to attempt to answer this question is with data collected directly on the input and output variables.

References

ASTM. (1996). *PS 64-96: Provisional Standard Guide for Developing Appropriate Statistical Approaches for Ground-Water Detection Monitoring Programs.* American Society for Testing and Materials, West Conshohocken, PA.

Aitchison, J. (1955). On the Distribution of a Positive Random Variable Having a Discrete Probability Mass at the Origin. *Journal of the American Statistical Association* **50**, 901–908.

Aitchison, J., and J.A.C. Brown (1957). *The Lognormal Distribution (with special references to its uses in economics).* Cambridge University Press, London, 176 pp.

Akritas, M., T. Ruscitti, and G.P. Patil. (1994). Statistical Analysis of Censored Environmental Data. In Patil, G.P., and C.R. Rao, eds., *Handbook of Statistics, Vol. 12: Environmental Statistics.* North-Holland, Amsterdam, Chapter 7, 221–242.

Amemiya, T. (1985). *Advanced Econometrics.* Harvard University Press, Cambridge, MA, 521 pp.

Arlinghaus, S.L., ed. (1996). *Practical Handbook of Spatial Statistics.* CRC Press, Boca Raton, FL, 307 pp.

Bain, L.J., and M. Engelhardt. (1991). *Statistical Analysis of Reliability and Life-Testing Models.* Marcel Dekker, New York, 496 pp.

Balakrishnan, N., and A.C. Cohen. (1991). *Order Statistics and Inference: Estimation Methods.* Academic Press, San Diego, CA, 375 pp.

Barclay's California Code of Regulations. (1991). Title 22, §66264.97 [concerning hazardous waste facilities] and Title 23, §2550.7(e)(8) [concerning solid waste facilities]. Barclay's Law Publishers, San Francisco, CA.

Barnett, V., and T. Lewis. (1995). *Outliers in Statistical Data.* Third Edition. John Wiley & Sons, New York, 584 pp.

Barnett, V., and A. O'Hagan. (1997). *Setting Environmental Standards: The Statistical Approach to Handling Uncertainty and Variation.* Chapman & Hall, London, 111 pp.

Barnett, V., and K.F. Turkman, eds. (1997). *Statistics for the Environment 3: Pollution Assessment and Control.* John Wiley & Sons, New York, 345 pp.

Barr, D.R., and T. Davidson. (1973). A Kolmogorov-Smirnov Test for Censored Samples. *Technometrics* **15**(4), 739–757.

Barry, T.M. (1996). Recommendations on the Testing and Use of Pseudo-Random Number Generators Used in Monte Carlo Analysis for Risk Assessment. *Risk Analysis* **16**, 93–105.

Bartels, R. (1982). The Rank Version of von Neumann's Ratio Test for Randomness. *Journal of the American Statistical Association* **77**(377), 40–46.

Berthouex, P.M., and I. Hau. (1991a). A Simple Rule for Judging Compliance Using Highly Censored Samples. *Research Journal of the Water Pollution Control Federation* **63**(6), 880–886.

Berthouex, P.M., and I. Hau. (1991b). Difficulties Related to Using Extreme Percentiles for Water Quality Regulations. *Research Journal of the Water Pollution Control Federation* **63**(6), 873–879.

Berthouex, P.M., and L.C. Brown. (1994). *Statistics for Environmental Engineers*. Lewis Publishers, Boca Raton, FL, 335 pp.

Birnbaum, Z. W., and F.H. Tingey. (1951). One-Sided Confidence Contours for Probability Distribution Functions. *Annals of Mathematical Statistics* **22**, 592–596.

Blom, G. (1958). *Statistical Estimates and Transformed Beta Variables*. John Wiley & Sons, New York.

Bogen, K.T. (1995). Methods to Approximate Joint Uncertainty and Variability in Risk. *Risk Analysis* **15**(3), 411–419.

Bose, A., and N. K. Neerchal. (1996). *Unbiased Estimation of Variance Using Ranked Set Sampling*. Technical Report # 96-11, Department of Statistics, Purdue University, West Lafayette, IN.

Boswell, M. T., and G. P. Patil. (1987). A Perspective of Composite Sampling. *Communications in Statistics—Theory and Methods* **16**, 3069–3093.

Box, G.E.P., and D.R. Cox. (1964). An Analysis of Transformations (with Discussion). *Journal of the Royal Statistical Society, Series B* **26**(2), 211–252.

Box, G.E.P., W.G. Hunter, and J.S. Hunter. (1978). *Statistics for Experimenters: An Introduction to Design, Data Analysis, and Model Building*. John Wiley & Sons, New York, 653 pp.

Box, G.E.P., and G.M. Jenkins. (1976). *Time Series Analysis: Forecasting and Control*. Prentice-Hall, Englewood Cliffs, NJ, 575 pp.

Bradu, D., and Y. Mundlak. (1970). Estimation in Lognormal Linear Models. *Journal of the American Statistical Association* **65**, 198–211.

Brainard, J., and D.E. Burmaster. (1992). Bivariate Distributions for Height and Weight of Men and Women in the United States. *Risk Analysis* **12**(2), 267–275.

Bukowski, J., L. Korn, and D. Wartenberg. (1995). Correlated Inputs in Quantitative Risk Assessment: The Effects of Distributional Shape. *Risk Analysis* **15**(2), 215–219.

Burmaster, D.E., and P.D. Anderson. (1994). Principles of Good Practice for the Use of Monte Carlo Techniques in Human Health and Ecological Risk Assessments. *Risk Analysis* **14**(4), 477–481.

Burmaster, D.E., and R.H. Harris. (1993). The Magnitude of Compounding Conservatisms in Superfund Risk Assessments. *Risk Analysis* **13**(2), 131–134.

Burmaster, D.E., and K.M. Thompson. (1997). Estimating Exposure Point Concentrations for Surface Soils for Use in Deterministic and Probabilistic Risk Assessments. *Human and Ecological Risk Assessment* **3**(3), 363–384.

Burmaster, D.E., and A.M. Wilson. (1996). An Introduction to Second-Order Random Variables in Human Health Risk Assessments. *Human and Ecological Risk Assessment* **2**(4), 892–919.

Byrnes, M.E. (1994). *Field Sampling Methods for Remedial Investigations.* Lewis Publishers, Boca Raton, FL, 254 pp.

Calitz, F. (1973). Maximum Likelihood Estimation of the Parameters of the Three-Parameter Lognormal Distribution—a Reconsideration. *Australian Journal of Statistics* **15**(3), 185–190.

Callahan, B.G., ed. (1998). Special Issue: Probabilistic Risk Assessment. *Human and Ecological Risk Assessment* **4**(2).

Callahan, B.G., D.E. Burmaster, R.L. Smith, D.D. Krewski, and B.A.F. DelloRusso, eds. (1996). Commemoration of the 50th Anniversary of Monte Carlo, *Human and Ecological Risk Assessment* **2**(4).

Castillo, E. (1988). *Extreme Value Theory in Engineering.* Academic Press, New York, 389 pp.

Castillo, E., and A. Hadi. (1994). Parameter and Quantile Estimation for the Generalized Extreme-Value Distribution. *Environmetrics* **5**, 417–432.

Caulcutt, R., and R. Boddy. (1983). *Statistics for Analytical Chemists.* Chapman & Hall, London, 253 pp.

Chambers, J.M., W.S. Cleveland, B. Kleiner, and P.A. Tukey. (1983). *Graphical Methods for Data Analysis.* Duxbury Press, Boston, MA, 395 pp.

Chen, J.J., D.W. Gaylor, and R.L. Kodell. (1990). Estimation of the Joint Risk from Multiple-Compound Exposure Based on Single-Compound Experiments. *Risk Analysis* **10**(2), 285–290.

Chen, L. (1995a). A Minimum Cost Estimator for the Mean of Positively Skewed Distributions with Applications to Estimation of Exposure to Contaminated Soils. *Environmetrics* **6**, 181–193.

Chen, L. (1995b). Testing the Mean of Skewed Distributions. *Journal of the American Statistical Association* **90**(430), 767–772.

Cheng, R.C.H., and N.A.K. Amin. (1982). Maximum Product-of-Spacings Estimation with Application to the Lognormal Distribution. *Journal of the Royal Statistical Society, Series B* **44**, 394–403.

Chew, V. (1968). Simultaneous Prediction Intervals. *Technometrics* **10**(2), 323–331.

Chou, Y.M., and D.B. Owen. (1986). One-sided Distribution-Free Simultaneous Prediction Limits for p Future Samples. *Journal of Quality Technology* **18**, 96–98.

Chowdhury, J.U., J.R. Stedinger, and L. H. Lu. (1991). Goodness-of-Fit Tests for Regional Generalized Extreme Value Flood Distributions. *Water Resources Research* **27**(7), 1765–1776.

Clark, M.J.R., and P.H. Whitfield. (1994). Conflicting Perspectives about Detection Limits and about the Censoring of Environmental Data. *Water Resources Bulletin* **30**(6), 1063–1079.

Clayton, C.A., J.W. Hines, and P.D. Elkins. (1987). Detection Limits with Specified Assurance Probabilities. *Analytical Chemistry* **59**, 2506–2514.

Cleveland, W.S. (1993). *Visualizing Data*. Hobart Press, Summit, NJ, 360 pp.

Cleveland, W.S. (1994). *The Elements of Graphing Data*. Revised Edition. Hobart Press, Summit, NJ, 297 pp.

Cochran, W.G. (1977). *Sampling Techniques*. Third Edition. John Wiley & Sons, New York, 428 pp.

Code of Federal Regulations. (1996). Definition and Procedure for the Determination of the Method Detection Limit—Revision 1.11. Title 40, Part 136, Appendix B, 7-1-96 Edition, pp. 265–267.

Cogliano, V.J. (1997). Plausible Upper Bounds: Are Their Sums Plausible? *Risk Analysis* **17**(1), 77–84.

Cohen, A.C. (1951). Estimating Parameters of Logarithmic-Normal Distributions by Maximum Likelihood. *Journal of the American Statistical Association* **46**, 206–212.

Cohen, A.C. (1959). Simplified Estimators for the Normal Distribution When Samples are Singly Censored or Truncated. *Technometrics* **1**(3), 217–237.

Cohen, A.C. (1963). Progressively Censored Samples in Life Testing. *Technometrics* **5**, 327–339.

Cohen, A.C. (1988). Three-Parameter Estimation. In Crow, E.L., and K. Shimizu, eds. *Lognormal Distributions: Theory and Applications*. Marcel Dekker, New York, Chapter 4.

Cohen, A.C. (1991). *Truncated and Censored Samples*. Marcel Dekker, New York, 312 pp.

Cohen, A.C., and B.J. Whitten. (1980). Estimation in the Three-Parameter Lognormal Distribution. *Journal of the American Statistical Association* **75**, 399–404.

Cohen, A.C., B.J. Whitten, and Y. Ding. (1985). Modified Moment Estimation for the Three-Parameter Lognormal Distribution. *Journal of Quality Technology* **17**, 92–99.

Cohen, M.A., and P.B. Ryan. (1989). Observations Less Than the Analytical Limit of Detection: A New Approach. *JAPCA* **39**(3), 328–329.

Cohn, T.A. (1988). *Adjusted Maximum Likelihood Estimation of the Moments of Lognormal Populations form Type I Censored Samples*. U.S. Geological Survey Open-File Report 88-350, 34 pp.

Cohn, T.A., L.L. DeLong, E.J. Gilroy, R.M. Hirsch, and D.K. Wells. (1989). Estimating Constituent Loads. *Water Resources Research* **25**(5), 937–942.

Cohn, T.A., E.J. Gilroy, and W.G. Baier. (1992). Estimating Fluvial Transport of Trace Constituents Using a Regression Model with Data Subject to Censoring. *Proceedings of the American Statistical Association, Section on Statistics and the Environment*.

Conover, W.J. (1980). *Practical Nonparametric Statistics*. Second Edition. John Wiley & Sons, New York, 493 pp.

Conover, W.J., M.E. Johnson, and M.M. Johnson. (1981). A Comparative Study of Tests for Homogeneity of Variances, with Applications to the Outer Continental Shelf Bidding Data. *Technometrics* **23**(4), 351–361.

Cothern, C.R. (1994). How Does Scientific Information in General and Statistical Information in Particular Input to the Environmental Regulatory Process? In Patil, G.P., and C.R. Rao, eds., *Handbook of Statistics, Vol. 12: Environmental Statistics*. North-Holland, Amsterdam, Chapter 25, 791–816.

Cothern, C.R. (1996). *Handbook for Environmental Risk Decision Making: Values, Perceptions, & Ethics*. Lewis Publishers, Boca Raton, FL, 408 pp.

Cothern, C.R., and N.P. Ross. (1994). *Environmental Statistics, Assessment, and Forecasting*. Lewis Publishers, Boca Raton, FL, 418 pp.

Cox, D.D., L.H. Cox, and K.B. Ensor. (1997). Spatial Sampling and the Environment: Some Issues and Directions. *Environmental and Ecological Statistics* **4**, 219–233.

Cox, D.R., and D.V. Hinkley. (1974). *Theoretical Statistics.* Chapman & Hall, New York, 511 pp.

Cox, L.H., and W.W. Piegorsch. (1996). Combining Environmental Information, I: Environmental Monitoring, Measurement and Assessment. *Environmetrics* **7**, 299–308.

Cressie, N.A.C. (1993). *Statistics for Spatial Data.* Revised Edition. John Wiley & Sons, New York, 900 pp.

Crow, E.L., and K. Shimizu, eds. (1988). *Lognormal Distributions: Theory and Applications.* Marcel Dekker, New York, 387 pp.

Csuros, M. (1997). *Environmental Sampling and Analysis, Lab Manual.* Lewis Publishers, Boca Raton, FL, 373 pp.

Cullen, A.C. (1994). Measures of Compounding Conservatism in Probabilistic Risk Assessment. *Risk Analysis* **14**(4), 389–393.

Cunnane, C. (1978). Unbiased Plotting Positions—A Review. *Journal of Hydrology* **37**(3/4), 205–222.

Currie, L.A. (1968). Limits for Qualitative Detection and Quantitative Determination: Application to Radiochemistry. *Annals of Chemistry* **40**, 586–593.

Currie, L.A. (1988). *Detection in Analytical Chemistry: Importance, Theory, and Practice.* American Chemical Society, Washington, D.C.

Currie, L.A. (1995). Nomenclature in Evaluation of Analytical Methods Including Detection and Quantification Capabilities. *Pure & Applied Chemistry* **67**(10), 1699–1723.

Currie, L.A. (1996). Foundations and Future of Detection and Quantification Limits. *Proceedings of the Section on Statistics and the Environment,* American Statistical Association, Alexandria, VA.

Currie, L.A. (1997). Detection: International Update, and Some Emerging Di-Lemmas Involving Calibration, the Blank, and Multiple Detection Decisions. *Chemometrics and Intelligent Laboratory Systems* **37**, 151–181.

D'Agostino, R.B. (1970). Transformation to Normality of the Null Distribution of g1. *Biometrika* **57**, 679–681.

D'Agostino, R.B. (1971). An Omnibus Test of Normality for Moderate and Large Size Samples. *Biometrika* **58**, 341–348.

D'Agostino, R.B. (1986a). Graphical Analysis. In D'Agostino, R.B., and M.A. Stephens, eds. *Goodness-of-Fit Techniques.* Marcel Dekker, New York, Chapter 2, pp. 7–62.

D'Agostino, R.B. (1986b). Tests for the Normal Distribution. In D'Agostino, R.B., and M.A. Stephens, eds. *Goodness-of-Fit Techniques*. Marcel Dekker, New York, Chapter 9, pp. 367–419.

D'Agostino, R.B., and E.S. Pearson (1973). Tests for Departures from Normality. Empirical Results for the Distributions of β2 and β1. *Biometrika* **60**(3), 613–622.

D'Agostino, R.B., and M.A. Stephens, eds. (1986). *Goodness-of-Fit Techniques*. Marcel Dekker, New York, 560 pp.

D'Agostino, R.B., and G.L. Tietjen (1973). Approaches to the Null Distribution of Öb1. *Biometrika* **60**(1), 169–173.

Dallal, G. E., and L. Wilkinson. (1986). An Analytic Approximation to the Distribution of Lilliefor's Test for Normality. *The American Statistician* **40**, 294–296.

Danziger, L., and S. Davis. (1964). Tables of Distribution-Free Tolerance Limits. *Annals of Mathematical Statistics* **35**(5), 1361–1365.

David, M. (1977). *Geostatistical Ore Reserve Estimation*. Elsevier, New York, 364 pp.

Davidson, J.R. (1994). *Elipgrid-PC: A PC Program for Calculating Hot Spot Probabilities*. ORNL/TM-12774. Oak Ridge National Laboratory, Oak Ridge, TN.

Davis, C.B. (1994). Environmental Regulatory Statistics. In Patil, G.P., and C.R. Rao, eds., *Handbook of Statistics, Vol. 12: Environmental Statistics*. North-Holland, Amsterdam, Chapter 26, 817–865.

Davis, C.B. (1997). Challenges in Regulatory Environmetrics. *Chemometrics and Intelligent Laboratory Systems* **37**, 43–53.

Davis, C.B. (1998a). *Ground-Water Statistics & Regulations: Principles, Progress and Problems*. Second Edition. Environmetrics & Statistics Limited, Henderson, NV.

Davis, C.B. (1998b). Personal Communication, September 3, 1998.

Davis, C.B. (1998c). *Power Comparisons for Control Chart and Prediction Limit Procedures*. EnviroStat Technical Report 98-1, Environmetrics & Statistics Limited, Henderson, NV.

Davis, C.B., and R.J. McNichols. (1987). One-sided Intervals for at Least p of m Observations from a Normal Population on Each of r Future Occasions. *Technometrics* **29**, 359–370.

Davis, C.B., and R.J. McNichols. (1994a). Ground Water Monitoring Statistics Update: Part I: Progress Since 1988. *Ground Water Monitoring and Remediation* **14**(4), 148–158.

Davis, C.B., and R.J. McNichols. (1994b). Ground Water Monitoring Statistics Update: Part II: Nonparametric Prediction Limits. *Ground Water Monitoring and Remediation* **14**(4), 159–175.

Davis, C.B., and R.J. McNichols. (1999). Simultaneous Nonparametric Prediction Limits (with Discusson). *Technometrics* **41**(2), 89–112.

Davis, C.S., and M.A. Stephens. (1978). Algorithm AS 128. Approximating the Covariance Matrix of Normal Order Statistics. *Applied Statistics* **27**, 206–212.

Diggle, P.J. (1983). *Statistical Analysis of Spatial Point Patterns*. Academic Press, London.

Downing, D.J., R.H. Gardner, and F.O. Hoffman. (1985). An Examination of Response-Surface Methodologies for Uncertainty Analysis in Assessment Models. *Technometrics* **27**(2), 151–163. See also Easterling, R.G. (1986). Discussion of Downing, Gardner, and Hoffman (1985). *Technometrics* **28**(1), 91–93.

Downton, F. (1966). Linear Estimates of Parameters in the Extreme Value Distribution. *Technometrics* **8**(1), 3–17.

Draper, N., and H. Smith. (1981). *Applied Regression Analysis*. Second Edition. John Wiley & Sons, New York, 709 pp.

Draper, N., and H. Smith. (1998). *Applied Regression Analysis*. Third Edition. John Wiley & Sons, New York, 706 pp.

Dunnett, C.W. (1955). A Multiple Comparisons Procedure for Comparing Several Treatments with a Control. *Journal of the American Statistical Association* **50**, 1096–1121.

Dunnett, C.W. (1964). New Tables for Multiple Comparisons with a Control. *Biometrics* **20**, 482–491.

Dunnett, C.W., and M. Sobel. (1955). Approximations to the Probability Integral and Certain Percentage Points of a Multivariate Analogue of Student's t-Distribution. *Biometrika* **42**, 258–260.

Edland, S.D., and G. van Belle. (1994). Decreased Sampling Costs and Improved Accuracy with Composite Sampling. In Cothern, C.R., and N.P. Ross, eds., *Environmental Statistics, Assessment, and Forecasting*, Lewis Publishers, Boca Raton, FL, Chapter 2, pp. 29–55.

Efron, B. (1979a). Bootstrap Methods: Another Look at the Jackknife. *Annals of Statistics* **7**, 1–26.

Efron, B. (1979b). Computers and the Theory of Statistics: Thinking the Unthinkable. *Society for Industrial and Applied Mathematics* **21**, 460–480.

Efron, B. (1981). Nonparametric Standard Errors and Confidence Intervals. *Canadian Journal of Statistics* **9**, 139–172.

Efron, B. (1982). *The Jackknife, the Bootstrap and Other Resampling Plans.* Society for Industrial and Applied Mathematics, Philadelphia, PA, 92 pp.

Efron, B., and R.J. Tibshirani. (1993). *An Introduction to the Bootstrap.* Chapman & Hall, New York, 436 pp.

Einax, J.W., H.W. Zwanziger, and S. Geib. (1997). *Chemometrics in Environmental Analysis.* VCH Publishers, John Wiley & Sons, New York, 384 pp.

Ellison, B.E. (1964). On Two-Sided Tolerance Intervals for a Normal Distribution. *Annals of Mathematical Statistics* **35**, 762–772.

El-Shaarawi, A.H. (1989). Inferences about the Mean from Censored Water Quality Data. *Water Resources Research* **25**(4) 685–690.

El-Shaarawi, A.H. (1993). Environmental Monitoring, Assessment, and Prediction of Change. *Environmetrics* **4**(4), 381–398.

El-Shaarawi, A.H., and D.M. Dolan. (1989). Maximum Likelihood Estimation of Water Quality Concentrations from Censored Data. *Canadian Journal of Fisheries and Aquatic Sciences* **46**, 1033–1039.

El-Shaarawi, A.H., and S.R. Esterby. (1992). Replacement of Censored Observations by a Constant: An Evaluation. *Water Research* **26**(6), 835–844.

El-Shaarawi, A.H., and A. Naderi. (1991). Statistical Inference from Multiply Censored Environmental Data. *Environmental Monitoring and Assessment* **17**, 339–347.

Eschenroeder, A.Q., and E.J. Faeder. (1988). A Monte Carlo Analysis of Health Risks from PCB-Contaminated Mineral Oil Transformer Fires. *Risk Analysis* **8**(2), 291–297.

Esterby, S.R. (1993). Trend Analysis Methods for Environmental Data. *Environmetrics* **4**(4), 459–481.

Evans, M., N. Hastings, and B. Peacock. (1993). *Statistical Distributions.* Second Edition. John Wiley & Sons, New York, 170 pp.

Everitt, B.S. (1998). *The Cambridge Dictionary of Statistics.* Cambridge University Press, New York, 360 pp.

Everitt, B.S. (1999). *Chance Rules: An Informal Guide to Probability, Risk, and Statistics.* Springer-Verlag, New York, 202 pp.

Fertig, K.W., and N.R. Mann. (1977). One-Sided Prediction Intervals for at Least p out of m Future Observations from a Normal Population. *Technometrics* **19**, 167–177.

Fill, H.D., and J.R. Stedinger. (1995). *L* Moment and Probability Plot Correlation Coefficient Goodness-of-Fit Tests for the Gumbel Distribution and Impact of Autocorrelation. *Water Resources Research* **31**(1), 225–229.

Filliben, J.J. (1975). The Probability Plot Correlation Coefficient Test for Normality. *Technometrics* **17**(1), 111–117.

Finley, B., and D. Paustenbach. (1994). The Benefits of Probabilistic Exposure Assessment: Three Case Studies Involving Contaminated Air, Water, and Soil. *Risk Analysis* **14**(1), 53–73.

Finley, B., D. Proctor, P. Scott, N. Harrington, D. Paustenback, and P. Price. (1994). Recommended Distributions for Exposure Factors Frequently Used in Health Risk Assessment. *Risk Analysis* **14**(4), 533–553.

Finney, D.J. (1941). On the Distribution of a Variate Whose Logarithm is Normally Distributed. *Supplement to the Journal of the Royal Statistical Society* **7**, 155–161.

Fisher, L.D., and G. van Belle. (1993). *Biostatistics: A Methodology for the Health Sciences*. John Wiley & Sons, New York, 991 pp.

Fleiss, J. L. (1981). *Statistical Methods for Rates and Proportions*. Second Edition. John Wiley & Sons, New York, 321 pp.

Fleming, T.R., and D.P. Harrington. (1981). A Class of Hypothesis Tests for One and Two Sample Censored Survival Data. *Communications in Statistics—Theory and Methods* **A10**(8), 763–794.

Fleming, T.R., and D.P. Harrington. (1991). *Counting Processes and Survival Analysis*. John Wiley & Sons, New York, 429 pp.

Fletcher, D.J., L. Kavalieris, and B.F.J. Manly, eds. (1998). *Statistics in Ecology and Environmental Monitoring 2: Decision Making and Risk Assessment in Biology*. Otago Conference Series No. 6, University of Otago Press, Dunedin, New Zealand, 220 pp.

Fletcher, D.J., and B.F.J. Manly, eds. (1994). *Statistics in Ecology and Environmental Monitoring*. Otago Conference Series No. 2. University of Otago Press, Dunedin, New Zealand, 269 pp.

Flora, J.D. (1991). Statistics and Environmental Regulations. *Environmetrics* **2**(2), 129–137.

Freudenburg, W.R., and J.A. Rursch. (1994). The Risks of "Putting the Numbers in Context": A Cautionary Tale. *Risk Analysis* **14**(6), 949–958.

Gaylor, D.W., and J.J. Chen. (1996). A Simple Upper Limit for the Sum of the Risks of the Components in a Mixture. *Risk Analysis* **16**(3), 395–398.

Gentle, J.E. (1985). Monte Carlo Methods. In Kotz, S., and N.L. Johnson, eds. *Encyclopedia of Statistics, Volume 5*. John Wiley & Sons, New York, 612–617.

Gibbons, R.D. (1987a). Statistical Prediction Intervals for the Evaluation of Ground-Water Quality. *Ground Water* **25**, 455–465.

Gibbons, R.D. (1987b). Statistical Models for the Analysis of Volatile Organic Compounds in Waste Disposal Sites. *Ground Water* **25**, 572–580.

Gibbons, R.D. (1990). A General Statistical Procedure for Ground-Water Detection Monitoring at Waste Disposal Facilities. *Ground Water* **28**, 235–243.

Gibbons, R.D. (1991a). Some Additional Nonparametric Prediction Limits for Ground-Water Detection Monitoring at Waste Disposal Facilities. *Ground Water* **29**, 729–736.

Gibbons, R.D. (1991b). Statistical Tolerance Limits for Ground-Water Monitoring. *Ground Water* **29**, 563–570.

Gibbons, R.D. (1994). *Statistical Methods for Groundwater Monitoring.* John Wiley & Sons, New York, 286 pp.

Gibbons, R.D. (1995). Some Statistical and Conceptual Issues in the Detection of Low-Level Environmental Pollutants (with Discussion). *Environmetrics* **2**, 125–167.

Gibbons, R.D. (1996). Some Conceptual and Statistical Issues in Analysis of Groundwater Monitoring Data. *Environmetrics* **7**, 185–199.

Gibbons, R.D., and J. Baker. (1991). The Properties of Various Statistical Prediction Intervals for Ground-Water Detection Monitoring. *Journal of Environmental Science and Health* **A26**(4), 535–553.

Gibbons, R.D., D.E. Coleman, and R.F. Maddalone. (1997a). An Alternative Minimum Level Definition for Analytical Quantification. *Environmental Science and Technology* **31**(7), 2071–2077. Comments and Discussion in Volume 31(12), 3727–3731, and Volume 32(15), 2346–2353.

Gibbons, R.D., D.E. Coleman, and R.F. Maddalone. (1997b). Response to Comment on "An Alternative Minimum Level Definition for Analytical Quantification." *Environmental Science and Technology* **31**(12), 3729–3731.

Gibbons, R.D., D.E. Coleman, and R.F. Maddalone. (1998). Response to Comment on "An Alternative Minimum Level Definition for Analytical Quantification." *Environmental Science and Technology* **32**(15), 2349–2353.

Gibbons, R.D., N.E. Grams, F.H. Jarke, and K.P. Stoub. (1992). Practical Quantitation Limits. *Chemometrics Intelligent Laboratory Systems* **12**, 225–235.

Gibbons, R.D., F.H. Jarke, and K.P. Stoub. (1991). Detection Limits: For Linear Calibration Curves with Increasing Variance and Multiple Future Detection Decisions. In Tatsch, D.E., ed. *Waste Testing and Quality Assurance: Volume 3.* American Society for Testing and Materials, Philadelphia, PA.

Gibrat, R. (1930). Une loi des repartition economiques: l'effect proportionnel. *Bull. Statist. Gen. Fr.* **19**, 469ff.

Gilbert, R.O. (1987). *Statistical Methods for Environmental Pollution Monitoring.* Van Nostrand Reinhold, New York, 320 pp.

Gilbert, R.O. (1994). An Overview of Statistical Issues Related to Environmental Cleanup. In Patil, G.P., and C.R. Rao, eds., *Handbook of Statistics, Vol. 12: Environmental Statistics.* North-Holland, Amsterdam, Chapter 27, 867–880.

Gilliom, R.J., and D.R. Helsel. (1986). Estimation of Distributional Parameters for Censored Trace Level Water Quality Data: 1. Estimation Techniques. *Water Resources Research* **22**, 135–146.

Gilliom, R.J., R.M. Hirsch, and E.J. Gilroy. (1984). Effect of Censoring Trace-Level Water-Quality Data on Trend-Detection Capability. *Environmental Science and Technology* **18**, 530–535.

Gilroy, E.J., R.M. Hirsch, and T.A. Cohn. (1990). Mean Square Error of Regression-Based Constituent Transport Estimates. *Water Resources Research* **26**(9), 2069–2077.

Ginevan, M.E., and D.E. Splitstone. (1997). Improving Remediation Decisions at Hazardous Waste Sites with Risk-Based Geostatistical Analysis. *Environmental Science and Technology* **31**(2), 92A–96A.

Glasser, J.A., D.L. Foerst, G.D. McKee, S.A. Quave, and W.L. Budde. (1981). Trace Analyses for Wastewaters. *Environmental Science and Technology* **15**, 1426–1435.

Gleit, A. (1985). Estimation for Small Normal Data Sets with Detection Limits. *Environmental Science and Technology* **19**, 1201–1206.

Graham, R.C. (1993). *Data Analysis for the Chemical Sciences: A Guide to Statistical Techniques.* VCH Publishers, New York, 536 pp. Distributed by John Wiley & Sons, New York.

Graham, S.L., K.J. Davis, W.H. Hansen, and C.H. Graham. (1975). Effects of Prolonged Ethylene Thiourea Ingestion on the Thyroid of the Rat. *Food and Cosmetics Toxicology* **13**(5), 493–499.

Greenwood, J.A., J.M. Landwehr, N.C. Matalas, and J.R. Wallis. (1979). Probability Weighted Moments: Definition and Relation to Parameters of Several Distributions Expressible in Inverse Form. *Water Resources Research* **15**(5), 1049–1054.

Griffiths, D.A. (1980). Interval Estimation for the Three-Parameter Lognormal Distribution via the Likelihood Function. *Applied Statistics* **29**, 58–68.

Gringorten, I.I. (1963). A Plotting Rule for Extreme Probability Paper. *Journal of Geophysical Research* **68**(3), 813–814.

Gupta, A.K. (1952). Estimation of the Mean and Standard Deviation of a Normal Population from a Censored Sample. *Biometrika* **39**, 260–273.

Guttman, I. (1970). *Statistical Tolerance Regions: Classical and Bayesian.* Hafner Publishing Co., Darien, CT, 150 pp.

Haas, C.N. (1997). Importance of Distributional Form in Characterizing Inputs to Monte Carlo Risk Assessments. *Risk Analysis* **17**(1), 107–113.

Haas, C.N., and P.A. Scheff. (1990). Estimation of Averages in Truncated Samples. *Environmental Science and Technology* **24**(6), 912–919.

Hahn, G., and W. Nelson. (1973). A Survey of Prediction Intervals and Their Applications. *Journal of Quality Technology* **5**, 178–188.

Hahn, G.J. (1969). Factors for Calculating Two-Sided Prediction Intervals for Samples from a Normal Distribution. *Journal of the American Statistical Association* **64**(327), 878–898.

Hahn, G.J. (1970a). Additional Factors for Calculating Prediction Intervals for Samples from a Normal Distribution. *Journal of the American Statistical Association* **65**(332), 1668–1676.

Hahn, G.J. (1970b). Statistical Intervals for a Normal Population, Part I: Tables, Examples and Applications. *Journal of Quality Technology* **2**(3), 115–125.

Hahn, G.J. (1970c). Statistical Intervals for a Normal Population, Part II: Formulas, Assumptions, Some Derivations. *Journal of Quality Technology* **2**(4), 195–206.

Hahn, G.J., and W.Q. Meeker. (1991). *Statistical Intervals: A Guide for Practitioners.* John Wiley & Sons, New York, 392 pp.

Haimes, Y.Y., T. Barry, and J.H. Lambert, eds. (1994). When and How Can You Specify a Probability Distribution When You Don't Know Much? *Risk Analysis* **14**(5), 661–706.

Hall, I.J., and R.R. Prairie. (1973). One-Sided Prediction Intervals to Contain at Least m out of k Future Observations. *Technometrics* **15**, 897–914.

Hall, I.J., R.R. Prairie, and C.K. Motlagh. (1975). Non-Parametric Prediction Intervals. *Journal of Quality Technology* **7**(3), 109–114.

Hallenbeck, W.H. (1993). *Quantitative Risk Assessment for Environmental and Occupational Health.* Second Edition. Lewis Publishers, Boca Raton, FL, 224 pp.

Hamby, D.M. (1994). A Review of Techniques for Parameter Sensitivity Analysis of Environmental Models. *Environmental Monitoring and Assessment* **32**, 135–154.

Hamed, M., and P.B. Bedient. (1997). On the Effect of Probability Distributions of Input Variables in Public Health Risk Assessment. *Risk Analysis* **17**(1), 97–105.

Harrington, D.P., and T.R. Fleming. (1982). A Class of Rank Test Procedures for Censored Survival Data. *Biometrika* **69**(3), 553–566.

Harter, H.L., and A.H. Moore. (1966). Local-Maximum-Likelihood Estimation of the Parameters of Three-Parameter Lognormal Populations from Complete and Censored Samples. *Journal of the American Statistical Association* **61**, 842–851.

Hashimoto, L.K., and R.R. Trussell. (1983). Evaluating Water Quality Data Near the Detection Limit. Paper presented at the Advanced Technology Conference, American Water Works Association, Las Vegas, Nevada, June 5–9, 1983.

Hastie, T.J., and R.J. Tibshirani. (1990). *Generalized Additive Models*. Chapman & Hall, New York.

Hattis, D. (1990). Three Candidate "Laws" of Uncertainty Analysis. *Risk Analysis* **10**(1), 11.

Hattis, D., and D.E. Burmaster. (1994). Assessment of Variability and Uncertainty Distributions for Practical Risk Analyses. *Risk Analysis* **14**(5), 713–730.

Hayes, B. (1993). The Wheel of Fortune. *American Scientist* **81**, 114–118.

Hazardous Materials Training and Research Institute (HMTRI). (1997). *Site Characterization: Sampling and Analysis*. Van Nostrand Reinhold, 316 pp. Distributed by John Wiley & Sons, New York.

Helsel, D.R. (1987). Advantages of Nonparametric Procedures for Analysis of Water Quality Data. *Hydrological Sciences Journal* **32**, 179–190.

Helsel, D.R. (1990). Less than Obvious: Statistical Treatment of Data Below the Detection Limit. *Environmental Science and Technology* **24**(12), 1766–1774.

Helsel, D.R., and T.A. Cohn. (1988). Estimation of Descriptive Statistics for Multiply Censored Water Quality Data. *Water Resources Research* **24**(12), 1997–2004.

Helsel, D.R., and R.J. Gilliom. (1986). Estimation of Distributional Parameters for Censored Trace Level Water Quality Data: 2. Verification and Applications. *Water Resources Research* **22**, 147–155.

Helsel, D.R., and R.M. Hirsch. (1988). Discussion of Applicability of the t-test for Detecting Trends in Water Quality Variables. *Water Resources Bulletin* **24**(1), 201–204.

Helsel, D.R., and R.M. Hirsch. (1992). *Statistical Methods in Water Resources Research*. Elsevier, New York, 522 pp.

Hettmansperger, T.P. (1984). *Statistical Inference Based on Ranks*. John Wiley & Sons, New York, 323 pp.

Heyde, C.C. (1963). On a Property of the Lognormal Distribution. *Journal of the Royal Statistical Society, Series B* **25**, 392–393.

Hill, B.M. (1963). The Three-Parameter Lognormal Distribution and Bayesian Analysis of a Point-Source Epidemic. *Journal of the American Statistical Association* **58**, 72–84.

Hinkley, D.V., and G. Runger. (1984). The Analysis of Transformed Data (with Discussion). *Journal of the American Statistical Association* **79**, 302–320.

Hirsch, R.M. (1988). Statistical Methods and Sampling Design for Estimating Step Trends in Surface-Water Quality. *Water Resources Bulletin* **24**(3), 493–503.

Hirsch, R.M., R.B. Alexander, and R.A. Smith. (1991). Selection of Methods for the Detection and Estimation of Trends in Water Quality. *Water Resources Research* **27**(5), 803–813.

Hirsch, R.M., and E.J. Gilroy. (1984). Methods of Fitting a Straight Line to Data: Examples in Water Resources. *Water Resources Bulletin* **20**(5), 705–711.

Hirsch, R.M., D.R. Helsel, T.A. Cohn, and E.J. Gilroy. (1993). Statistical Analysis of Hydrologic Data. In Maidment, D.R., ed. *Handbook of Hydrology*. McGraw-Hill, New York, Chapter 17.

Hirsch, R.M., and J.R. Slack. (1984). A Nonparametric Trend Test for Seasonal Data with Serial Dependence. *Water Resources Research* **20**(6), 727–732.

Hirsch, R.M., J.R. Slack, and R.A. Smith. (1982). Techniques of Trend Analysis for Monthly Water Quality Data. *Water Resources Research* **18**(1), 107–121.

Hirsch, R.M., and J.R. Stedinger. (1987). Plotting Positions for Historical Floods and Their Precision. *Water Resources Research* **23**(4), 715–727.

Hoaglin, D.C. (1988). Transformations in Everyday Experience. *Chance* **1**, 40–45.

Hoaglin, D.C., and D.F. Andrews. (1975). The Reporting of Computation-Based Results in Statistics. *The American Statistician* **29**(3), 122–126.

Hoaglin, D.C., F. Mosteller, and J.W. Tukey, eds. (1983). *Understanding Robust and Exploratory Data Analysis*. John Wiley & Sons, New York, 447 pp.

Hoeffding, W. (1948). A Class of Statistics with Asymptotically Normal Distribution. *Annals of Mathematical Statistics* **19**, 293–325.

Hoffman, F.O., and J.S. Hammonds. (1994). Propagation of Uncertainty in Risk Assessments: The Need to Distinguish between Uncertainty Due to Lack of Knowledge and Uncertainty Due to Variability. *Risk Analysis* **14**(5), 707–712.

Hollander, M., and D.A. Wolfe. (1999). *Nonparametric Statistical Methods.* Second Edition. John Wiley & Sons, New York, 787 pp.

Hoshi, K., J.R. Stedinger, and J. Burges. (1984). Estimation of Log-Normal Quantiles: Monte Carlo Results and First-Order Approximations. *Journal of Hydrology* **71**, 1–30.

Hosking, J.R.M. (1984). Testing Whether the Shape Parameter is Zero in the Generalized Extreme-Value Distribution. *Biometrika* **71**(2), 367–374.

Hosking, J.R.M. (1985). Algorithm AS 215: Maximum-Likelihood Estimation of the Parameters of the Generalized Extreme-Value Distribution. *Applied Statistics* **34**(3), 301–310.

Hosking, J.R.M. (1990). L-Moments: Analysis and Estimation of Distributions Using Linear Combinations of Order Statistics. *Journal of the Royal Statistical Society, Series B* **52**(1), 105–124.

Hosking, J.R.M., and J.R. Wallis (1995). A Comparison of Unbiased and Plotting-Position Estimators of L Moments. *Water Resources Research* **31**(8), 2019–2025.

Hosking, J.R.M., J.R. Wallis, and E.F. Wood. (1985). Estimation of the Generalized Extreme-Value Distribution by the Method of Probability-Weighted Moments. *Technometrics* **27**(3), 251–261.

Hubaux, A., and G. Vos. (1970). Decision and Detection Limits for Linear Calibration Curves. *Annals of Chemistry* **42**, 849–855.

Hughes, J.P., and S.P. Millard. (1988). A Tau-Like Test for Trend in the Presence of Multiple Censoring Points. *Water Resources Bulletin* **24**(3), 521–531.

Ibrekk, H., and M.G. Morgan. (1987). Graphical Communication of Uncertain Quantities to Nontechnical People. *Risk Analysis* **7**(4), 519–529.

Iman, R.L., and W.J. Conover. (1980). Small Sample Sensitivity Analysis Techniques for Computer Models, With an Application to Risk Assessment (with Comments). *Communications in Statistics—Volume A, Theory and Methods* **9**(17), 1749–1874.

Iman, R.L., and W.J. Conover. (1982). A Distribution-Free Approach to Inducing Rank Correlation Among Input Variables. *Communications in Statistics—Volume B, Simulation and Computation* **11**(3), 311–334.

Iman, R.L., and J.M. Davenport. (1982). Rank Correlation Plots For Use with Correlated Input Variables. *Communications in Statistics—Volume B, Simulation and Computation* **11**(3), 335–360.

Iman, R.L., and J.C. Helton. (1988). An Investigation of Uncertainty and Sensitivity Analysis Techniques for Computer Models. *Risk Analysis* **8**(1), 71–90.

Iman, R.L., and J.C. Helton. (1991). The Repeatability of Uncertainty and Sensitivity Analyses for Complex Probabilistic Risk Assessments. *Risk Analysis* **11**(4), 591–606.

Isaaks, E.H., and R.M. Srivastava. (1989). *An Introduction to Applied Geostatistics*. Oxford University Press, New York, 561 pp.

Israeli, M., and C.B. Nelson. (1992). Distribution and Expected Time of Residence for U.S. Households. *Risk Analysis* **12**(1), 65–72.

Jenkinson, A.F. (1955). The Frequency Distribution of the Annual Maximum (or Minimum) of Meteorological Events. *Quarterly Journal of the Royal Meteorological Society* **81**, 158–171.

Jenkinson, A.F. (1969). *Statistics of Extremes*. Technical Note 98, World Meteorological Office, Geneva.

Johnson, G.D., B.D. Nussbaum, G.P. Patil, and N.P. Ross. (1996). Designing Cost-Effective Environmental Sampling Using Concomitant Information. *Chance* **9**(1), 4–11.

Johnson, N.L., S. Kotz, and N. Balakrishnan. (1994). *Continuous Univariate Distributions, Volume 1*. Second Edition. John Wiley & Sons, New York, 756 pp.

Johnson, N.L., S. Kotz, and N. Balakrishnan. (1995). *Continuous Univariate Distributions, Volume 2*. Second Edition. John Wiley & Sons, New York, 719 pp.

Johnson, N.L., S. Kotz, and N. Balakrishnan. (1997). *Discrete Multivariate Distributions*. John Wiley & Sons, New York, 299 pp.

Johnson, N.L., S. Kotz, and A.W. Kemp. (1992). *Univariate Discrete Distributions*. Second Edition. John Wiley & Sons, New York, 565 pp.

Johnson, N.L., and B.L. Welch. (1940). Applications of the Non-Central t-Distribution. *Biometrika* **31**, 362–389.

Johnson, R.A., D.R. Gan, and P.M. Berthouex. (1995). Goodness-of-Fit Using Very Small but Related Samples with Application to Censored Data Estimation of PCB Contamination. *Environmetrics* **6**, 341–348.

Johnson, R.A., S. Verrill, and D.H. Moore. (1987). Two-Sample Rank Tests for Detecting Changes That Occur in a Small Proportion of the Treated Population. *Biometrics* **43**, 641–655.

Johnson, R.A., and D.W. Wichern. (1998). *Applied Multivariate Statistical Analysis*. Fourth Edition. Prentice-Hall, Englewood Cliffs, NJ, 816 pp.

Journel, A.G., and Huijbregts, C.J. (1978). *Mining Geostatistics*. Academic Press, London.

Judge, G.G., W.E. Griffiths, R.C. Hill, H. Lutkepohl, and T.C. Lee. (1985). Qualitative and Limited Dependent Variable Models: Chapter 18. In *The Theory and Practice of Econometrics*. John Wiley & Sons, New York, 1019 pp.

Kahn, H.D., W.A. Telliard, and C.E. White. (1998). Comment on "An Alternative Minimum Level Definition for Analytical Quantification" (with Response). *Environmental Science and Technology* **32**(5), 2346–2353.

Kahn, H.D., C.E. White, K. Stralka, and R. Kuznetsovski. (1997). Alternative Estimates of Detection. *Proceedings of the Twentieth Annual EPA Conference on Analysis of Pollutants in the Environment*, May 7–8, Norfolk, VA. U.S. Environmental Protection Agency, Washington, D.C.

Kaiser, H. (1965). Zum Problem der Nachweisgrenze. *Fresenius' Z. Anal. Chem.* **209**, 1.

Kalbfleisch, J.D., and R.L. Prentice. (1980). *The Statistical Analysis of Failure Time Data*. John Wiley & Sons, New York, 321 pp.

Kaluzny, S.P., S.C. Vega, T.P. Cardosa, and A.A. Shelly. (1998). *S+SPATIALSTATS User's Manual for Windows and UNIX*. Springer-Verlag, New York, 327 pp.

Kaplan, E.L., and P. Meier. (1958). Nonparametric Estimation from Incomplete Observations. *Journal of the American Statistical Association* **53**, 457–481.

Kapteyn, J.C. (1903). *Skew Frequency Curves in Biology and Statistics*. Astronomical Laboratory, Noordhoff, Groningen.

Keith, L.H. (1991). *Environmental Sampling and Analysis: A Practical Guide*. Lewis Publishers, Boca Raton, FL, 143 pp.

Keith, L.H., ed. (1996). *Principles of Environmental Sampling*. Second Edition. American Chemical Society, Washington, D.C., 848 pp. Distributed by Oxford University Press, New York.

Kempthorne, O., and T.E. Doerfler. (1969). The Behavior of Some Significance Tests under Experimental Randomization. *Biometrika* **56**(2), 231–248.

Kendall, M.G., and J.D. Gibbons. (1990). *Rank Correlation Methods*. Fifth Edition. Charles Griffin, London, 260 pp.

Kendall, M.G., and A. Stuart. (1991). *The Advanced Theory of Statistics, Volume 2: Inference and Relationship*. Fifth Edition. Oxford University Press, New York, 1323 pp.

Kennedy, W.J., and J.E. Gentle. (1980). *Statistical Computing*. Marcel Dekker, New York, 591 pp.

Kim, P.J., and R.I. Jennrich. (1973). Tables of the Exact Sampling Distribution of the Two Sample Kolmogorov-Smirnov Criterion. In Harter, H.L., and D.B. Owen, eds. *Selected Tables in Mathematical Statistics, Vol. 1*. American Mathematical Society, Providence, RI, pp. 79–170.

Kimbrough, D.E. (1997). Comment on "An Alternative Minimum Level Definition for Analytical Quantification" (with Response). *Environmental Science and Technology* **31**(12), 3727–3731.

Kitanidis, P.K. (1997). *Introduction to Geostatistics*. Cambridge University Press, London, 249 pp.

Kodell, R.L., H. Ahn, J.J. Chen, J.A. Springer, C.N. Barton, and R.C. Hertzberg. (1995). Upper Bound Risk Estimates for Mixtures of Carcinogens. *Toxicology* **105**, 199–208.

Kodell, R.L., and J.J. Chen. (1994). Reducing Conservatism in Risk Estimation for Mixtures of Carcinogens. *Risk Analysis* **14**(3), 327–332.

Kolmogorov, A.N. (1933). Sulla determinazione empirica di una legge di distribuzione. *Giornale dell' Istituto Italiano degle Attuari* **4**, 83–91.

Korn, L.R., and D.E. Tyler. (2001). Robust Estimation for Chemical Concentration Data Subject to Detection Limits. In Fernholz, L., S. Morgenthaler, and W. Stahel, eds. *Statistics in Genetics and in the Environmental Sciences*. Birkhauser Verlag, Basel, pp.41-63.

Kotz, S., and N.L. Johnson, eds. (1985). *Encyclopedia of Statistics*. John Wiley & Sons, New York.

Kuchenhoff, H., and M. Thamerus. (1996). Extreme Value Analysis of Munich Air Pollution Data. *Environmental and Ecological Statistics* **3**, 127–141.

Kushner, E.J. (1976). On Determining the Statistical Parameters for Pollution Concentrations from a Truncated Data Set. *Atmospheric Environment* **10**, 975–979.

Lambert, D., B. Peterson, and I. Terpenning. (1991). Nondetects, Detection Limits, and the Probability of Detection. *Journal of the American Statistical Association* **86**(414), 266–277.

Lambert, J.H., N.C. Matalas, C.W. Ling, Y.Y. Haimes, and D. Li. (1994). Selection of Probability Distributions in Characterizing Risk of Extreme Events. *Risk Analysis* **14**(5), 731–742.

Land, C.E. (1971). Confidence Intervals for Linear Functions of the Normal Mean and Variance. *The Annals of Mathematical Statistics* **42**(4), 1187–1205.

Land, C.E. (1972). An Evaluation of Approximate Confidence Interval Estimation Methods for Lognormal Means. *Technometrics* **14**(1), 145–158.

Land, C.E. (1973). Standard Confidence Limits for Linear Functions of the Normal Mean and Variance. *Journal of the American Statistical Association* **68**(344), 960–963.

Land, C.E. (1975). Tables of confidence limits for linear functions of the normal mean and variance, in *Selected Tables in Mathematical Statistics, Vol. III.* American Mathematical Society, Providence, RI, pp. 385–419.

Landwehr, J.M., N.C. Matalas, and J.R. Wallis. (1979). Probability Weighted Moments Compared with Some Traditional Techniques in Estimating Gumbel Parameters and Quantiles. *Water Resources Research* **15**(5), 1055–1064.

Laudan, L. (1997). *Danger Ahead: The Risks You Really Face on Life's Highway.* John Wiley & Sons, New York, 203 pp.

Law, A.M., and W.D. Kelton. (1991). *Simulation Modeling & Analysis.* Second Edition. McGraw-Hill, Inc., New York, 759 pp.

Latta, R.B. (1981). A Monte Carlo Study of Some Two-Sample Rank Tests with Censored Data. *Journal of the American Statistical Association* **76**(375), 713–719.

Lee, E.T. (1980). *Statistical Methods for Survival Data Analysis.* Lifetime Learning Publications, Belmont, CA, 557 pp.

Lehmann, E.L. (1975). *Nonparametrics: Statistical Methods Based on Ranks.* Holden-Day, Oakland, CA, 457 pp.

Leidel, N.A., K.A. Busch, and J.R. Lynch. (1977). *Occupational Exposure Sampling Strategy Manual.* U.S. Department of Health, Education, and Welfare, Public Health Service, Center for Disease Control, National Institute for Occupational Safety and Health, Cincinnati, OH, January, 1977.

Lewis, H.W. (1990). *Technological Risk.* W.W. Norton & Company, New York, 353 pp.

Lettenmaier, D.P. (1976). Detection of Trends in Water Quality Data from Data with Dependent Observations. *Water Resources Research* **12**, 1037–1046.

Lettenmaier, D.P. (1988). Multivariate Non-Parametric Tests for Trend in Water Quality. *Water Resources Bulletin* **24**(3), 505–512.

Likes, J. (1980). Variance of the MVUE for Lognormal Variance. *Technometrics* **22**(2), 253–258.

Looney, S.W., and T.R. Gulledge. (1985). Use of the Correlation Coefficient with Normal Probability Plots. *The American Statistician* **39**(1), 75–79.

Lovison, G., S.D. Gore, and G.P. Patil. (1994). Design and Analysis of Composite Sampling Procedures: A Review. In Patil, G.P., and C.R. Rao, eds., *Handbook of Statistics, Vol. 12: Environmental Statistics*. North-Holland, Amsterdam, Chapter 4, 103–166.

Lu, L., and J.R. Stedinger. (1992). Variance of Two- and Three-Parameter GEV/PWM Quantile Estimators: Formulae, Confidence Intervals, and a Comparison. *Journal of Hydrology* **138**, 247–267.

Lucas, J.M. (1982). Combined Shewart-CUMSUM Quality Control Schemes. *Journal of Quality Technology* **14**, 51–59.

Lucas, J.M. (1985). Cumulative Sum (CUMSUM) Control Schemes. *Communications in Statistics A—Theory and Methods* **14**(11), 2689–2704.

Lundgren, R., and A. McMakin. (1998). *Risk Communication: A Handbook for Communicating Environmental, Safety, and Health Risks*. Battelle Press, Columbus, OH, 362 pp.

Macleod, A.J. (1989). Remark AS R76: A Remark on Algorithm AS 215: Maximum Likelihood Estimation of the Parameters of the Generalized Extreme-Value Distribution. *Applied Statistics* **38**(1), 198–199.

Manly, B.F.J. (1997). Randomization, Bootstrap and Monte Carlo Methods in Biology. Second Edition. Chapman & Hall, New York, 399 pp.

Maritz, J.S., and A.H. Munro. (1967). On the Use of the Generalized Extreme-Value Distribution in Estimating Extreme Percentiles. *Biometrics* **23**(1), 79–103.

Marsaglia, G. et al. (1973). *Random Number Package: "Super-Duper."* School of Computer Science, McGill University, Montreal.

Massart, D.L., B.G.M. Vandeginste, S.N. Deming, Y. Michotte, and L. Kaufman. (1988). *Chemometrics: A Textbook*. Elsevier, New York, Chapter 7.

McBean, E.A., and F.A. Rovers. (1992). Estimation of the Probability of Exceedance of Contaminant Concentrations. *Ground Water Monitoring Review* **Winter**, 115–119.

McBean, E.A., and R.A. Rovers. (1998). *Statistical Procedures for Analysis of Environmental Monitoring Data & Risk Assessment*. Prentice-Hall PTR, Upper Saddle River, NJ, 313 pp.

McCullagh, P., and J.A. Nelder. (1989). *Generalized Linear Models*. Second Edition. Chapman & Hall, New York, 511 pp.

McIntyre, G.A. (1952). A Method of Unbiased Selective Sampling, Using Ranked Sets. *Australian Journal of Agricultural Resources* **3**, 385–390.

McKay, M.D., R.J. Beckman., and W.J. Conover. (1979). A Comparison of Three Methods for Selecting Values of Input Variables in the Analysis of Output from a Computer Code. *Technometrics* **21**(2), 239–245.

McKean, J.W., and G.L. Sievers. (1989). Rank Scores Suitable for Analysis of Linear Models under Asymmetric Error Distributions. *Technometrics* **31**, 207–218.

McKone, T.E. (1994). Uncertainty and Variability in Human Exposures to Soil and Contaminants through Home-Grown Food: A Monte Carlo Assessment. *Risk Analysis* **14**(4), 449–463.

McKone, T.E., and K.T. Bogen. (1991). Predicting the Uncertainties in Risk Assessment. *Environmental Science and Technology* **25**(10), 1674–1681.

McLeod, A.I., K.W. Hipel, and B.A. Bodo. (1991). Trend Analysis Methodology for Water Quality Time Series. *Environmetrics* **2**(2), 169–200.

McNichols, R.J., and C.B. Davis. (1988). Statistical Issues and Problems in Ground Water Detection Monitoring at Hazardous Waste Facilities. *Ground Water Monitoring Review* **8**(4), 135–150.

Meeker, W.Q., and L.A. Escobar. (1998). *Statistical Methods for Reliability Data*. John Wiley & Sons, New York.

Meier, P.C., and R.E. Zund. (1993). *Statistical Methods in Analytical Chemistry*. John Wiley & Sons, New York, 321 pp.

Michael, J.R., and W.R. Schucany. (1986). Analysis of Data from Censored Samples. In D'Agostino, R.B., and M.A. Stephens, eds. *Goodness-of-Fit Techniques*. Marcel Dekker, New York, 560 pp., Chapter 11, 461–496.

Millard, S.P. (1987a). Environmental Monitoring, Statistics, and the Law: Room for Improvement (with Comment). *The American Statistician* **41**(4), 249–259.

Millard, S.P. (1987b). Proof of Safety vs. Proof of Hazard. *Biometrics* **43**, 719–725.

Millard, S.P. (1996). Estimating a Percent Reduction in Load. *Water Resources Research* **32**(6), 1761–1766.

Millard, S.P. (1998). *ENVIRONMENTALSTATS for S-PLUS User's Manual for Windows and UNIX*. Springer-Verlag, New York, 381 pp.

Millard, S.P., and S.J. Deverel. (1988). Nonparametric Statistical Methods for Comparing Two Sites Based on Data with Multiple Nondetect Limits. *Water Resources Research* **24**(12), 2087–2098.

Millard, S.P., and D.P. Lettenmaier. (1986). Optimal Design of Biological Sampling Programs Using the Analysis of Variance. *Estuarine, Coastal and Shelf Science* **22**, 637–656.

Millard, S.P., and N.K. Neerchal. (2001). *Environmental Statistics with S-PLUS.* CRC Press, Boca Raton, Florida, 830 pp.

Millard, S.P., J.R. Yearsley, and D.P. Lettenmaier. (1985). Space-Time Correlation and Its Effects on Methods for Detecting Aquatic Ecological Change. *Canadian Journal of Fisheries and Aquatic Science,* **42**(8), 1391–1400. Correction: (1986), 43, 1680.

Miller, R.G. (1981a). *Simultaneous Statistical Inference.* Springer-Verlag, New York, 299 pp.

Miller, R.G. (1981b). *Survival Analysis.* John Wiley & Sons, New York, 238 pp.

Milliken, G.A., and D.E. Johnson. (1992). *Analysis of Messy Data, Volume I: Designed Experiments.* Chapman & Hall, New York, 473 pp.

Mode, N.A., L.L. Conquest, and D.A. Marker. (1999). Ranked Set Sampling for Ecological Research: Accounting for the Total Costs of Sampling. *Environmetrics* **10**, 179–194.

Molak, V., ed. (1997). *Fundamentals of Risk Analysis and Risk Management.* Lewis Publishers, Boca Raton, FL, 472 pp.

Montgomery, D.C. (1997). *Introduction to Statistical Quality Control.* Third Edition. John Wiley & Sons, New York.

Moore, D.S. (1986). Tests of Chi-Squared Type. In D'Agostino, R.B., and M.A. Stephens, eds. *Goodness-of-Fit Techniques.* Marcel Dekker, New York, pp. 63–95.

Morgan, M.G., and M. Henrion. (1990). *Uncertainty: A Guide to Dealing with Uncertainty in Quantitative Risk and Policy Analysis.* Cambridge University Press, New York, 332 pp.

Neely, W.B. (1994). *Introduction to Chemical Exposure and Risk Assessment.* Lewis Publishers, Boca Raton, FL, 190 pp.

Neerchal, N. K., and S. L. Brunenmeister. (1993). Estimation of Trend in Chesapeake Bay Water Quality Data. In Patil, G.P., and C.R. Rao, eds., *Handbook of Statistics, Vol. 6: Multivariate Environmental Statistics.* North-Holland, Amsterdam, Chapter 19, 407–422.

Nelson, W. (1969). Hazard Plotting for Incomplete Failure Data. *Journal of Quality Technology* **1**(1), 27–52.

Nelson, W. (1972). Theory and Applications of Hazard Plotting for Censored Failure Data. *Technometrics* **14**, 945–966.

Nelson, W. (1982). *Applied Life Data Analysis.* John Wiley & Sons, New York, 634 pp.

Nelson, W., and J. Schmee. (1979). Inference for (Log) Normal Life Distributions from Small Singly Censored Samples and BLUEs. *Technometrics* **21**, 43–54.

Nelson, W.R. (1970). Confidence Intervals for the Ratio of Two Poisson Means and Poisson Predictor Intervals. *IEEE Transactions on Reliability* **R-19**(2), 42–49.

Neptune, D., E.P. Brantly, M.J. Messner, and D.I. Michael. (1990). Quantitative Decision Making in Superfund: A Data Quality Objectives Case Study. *Hazardous Material Control* **May/June**, pp. 18–27.

Newman, M.C. (1995). *Quantitative Methods in Aquatic Ecotoxicology.* Lewis Publishers, Boca Raton, FL, 426 pp.

Newman, M.C., and P.M. Dixon. (1990). UNCENSOR: A Program to Estimate Means and Standard Deviations for Data Sets with Below Detection Limit Observations. *American Environmental Laboratory* **4(90)**, 26–30.

Newman, M.C., P.M. Dixon, B.B. Looney, and J.E. Pinder. (1989). Estimating Mean and Variance for Environmental Samples with Below Detection Limit Observations. *Water Resources Bulletin* **25**(4), 905–916.

Odeh, R.E., and D.B. Owen. (1980). *Tables for Normal Tolerance Limits, Sampling Plans, and Screening.* Marcel Dekker, New York, 316 pp.

Osborne, C. (1991). Statistical Calibration: A Review. *International Statistical Review* **59**(3), 309–336.

Ott, W.R. (1990). A Physical Explanation of the Lognormality of Pollutant Concentrations. *Journal of the Air and Waste Management Association* **40**, 1378–1383.

Ott, W.R. (1995). *Environmental Statistics and Data Analysis.* Lewis Publishers, Boca Raton, FL, 313 pp.

Owen, D.B. (1962). *Handbook of Statistical Tables.* Addison-Wesley, Reading, MA, 580 pp.

Owen, W., and T. DeRouen. (1980). Estimation of the Mean for Lognormal Data Containing Zeros and Left-Censored Values, with Applications to the Measurement of Worker Exposure to Air Contaminants. *Biometrics* **36**, 707–719.

Parkin, T.B., S.T. Chester, and J.A. Robinson. (1990). Calculating Confidence Intervals for the Mean of a Lognormally Distributed Variable. *Journal of the Soil Science Society of America* **54**, 321–326.

Parkin, T.B., J.J. Meisinger, S.T. Chester, J.L. Starr, and J.A. Robinson. (1988). Evaluation of Statistical Estimation Methods for Lognormally Distributed Variables. *Journal of the Soil Science Society of America* **52**, 323–329.

Parmar, M.K.B., and D. Machin. (1995). *Survival Analysis: A Practical Approach.* John Wiley & Sons, New York, 255 pp.

Patil, G.P., S.D. Gore, and A.K. Sinha. (1994). Environmental Chemistry, Statistical Modeling, and Observational Economy. In Cothern, C.R., and N.P. Ross, eds., *Environmental Statistics, Assessment, and Forecasting*, Lewis Publishers, Boca Raton, FL, Chapter 3, pp. 57–97.

Patil, G.P., and C.R. Rao, eds. (1994). *Handbook of Statistics, Vol. 12: Environmental Statistics.* North-Holland, Amsterdam, 927 pp.

Patil, G.P., A.K. Sinha, and C. Taillie. (1994). Ranked Set Sampling. In Patil, G.P., and C.R. Rao, eds., *Handbook of Statistics, Vol. 12: Environmental Statistics.* North-Holland, Amsterdam, Chapter 5, 167–200.

Pearson, E.S., and H.O. Hartley, eds. (1970). *Biometrika Tables for Statisticians, Volume 1.* Cambridge University Press, New York, 270 pp.

Persson, T. and H. Rootzen. (1977). Simple and Highly Efficient Estimators for a Type I Censored Normal Sample. *Biometrika* **64**, 123–128.

Peterson, B.P., S.P. Millard, E.F. Wood, and D.P. Lettenmaier. (1990). Design of a Soil Sampling Study to Determine the Habitability of the Emergency Declaration Area, Love Canal, New York. *Environmetrics* **1**(1), 89–119.

Pettitt, A.N. (1976). Cramér-von Mises Statistics for Testing Normality with Censored Samples. *Biometrika* **63**(3), 475–481.

Pettitt, A. N. (1983). Re-Weighted Least Squares Estimation with Censored and Grouped Data: An Application of the EM Algorithm. *Journal of the Royal Statistical Society, Series B* **47**, 253-260.

Pettitt, A.N., and M.A. Stephens. (1976). Modified Cramér-von Mises Statistics for Censored Data. *Biometrika* **63**(2), 291–298.

Piegorsch, W.W., and A.J. Bailer. (1997). *Statistics for Environmental Biology and Toxicology.* Chapman & Hall, New York, 579 pp.

Pomeranz, J. (1973). Exact Cumulative Distribution of the Kolmogorov-Smirnov Statistic for Small Samples (Algorithm 487). *Collected Algorithms from CACM.*

Porter, P.S., R.C. Ward, and H.F. Bell. (1988). The Detection Limit. *Environmental Science and Technology* **22**(8), 856–861.

Prentice, R.L. (1978). Linear Rank Tests with Right Censored Data. *Biometrika* **65**, 167–179.

Prentice, R.L. (1985). Linear Rank Tests. In Kotz, S., and N.L. Johnson, eds. *Encyclopedia of Statistical Science.* John Wiley & Sons, New York. Volume 5, pp. 51–58.

Prentice, R.L., and P. Marek. (1979). A Qualitative Discrepancy between Censored Data Rank Tests. *Biometrics* **35**, 861–867.

Prescott, P., and A.T. Walden. (1980). Maximum Likelihood Estimation of the Parameters of the Generalized Extreme-Value Distribution. *Biometrika* **67**(3), 723–724.

Prescott, P., and A.T. Walden. (1983). Maximum Likelihood Estimation of the Three-Parameter Generalized Extreme-Value Distribution from Censored Samples. *Journal of Statistical Computing and Simulation* **16**, 241–250.

Press, W.H., S.A. Teukolsky, W.T. Vetterling, and B.P. Flannery. (1992). *Numerical Recipes in FORTRAN: The Art of Scientific Computing.* Second Edition. Cambridge University Press, New York, 963 pp.

Qian, S.S. (1997). Estimating the Area Affected by Phosphorus Runoff in an Everglades Wetland: A Comparison of Universal Kriging and Bayesian Kriging. *Environmental and Ecological Statistics* **4**, 1–29.

Ranasinghe, J.A., L.C. Scott, and R. Newport. (1992). *Long-term Benthic Monitoring and Assessment Program for the Maryland Portion of the Bay, Jul 1984-Dec 1991.* Report prepared for the Maryland Department of the Environment and the Maryland Department of Natural Resources by Versar, Inc., Columbia, MD.

Ripley, B.D. (1981). *Spatial Statistics.* John Wiley & Sons, New York.

Rocke, D.M., and S. Lorenzato. (1995). A Two-Component Model for Measurement Error in Analytical Chemistry. *Technometrics* **37**(2), 176–184.

Rodricks, J.V. (1992). *Calculated Risks: The Toxicity and Human Health Risks of Chemicals in Our Environment.* Cambridge University Press, New York, 256 pp.

Rogers, J. (1992). Assessing Attainment of Ground Water Cleanup Standards Using Modified Sequential t-Tests. *Environmetrics* **3**(3), 335–359.

Rosebury, A.M., and D.E. Burmaster. (1992). Lognormal Distributions for Water Intake by Children and Adults. *Risk Analysis* **12**(1), 99–104.

Rowe, W.D. (1994). Understanding Uncertainty. *Risk Analysis* **14**(5), 743–750.

Royston, J.P. (1982). Algorithm AS 177. Expected Normal Order Statistics (Exact and Approximate). *Applied Statistics* **31**, 161–165.

Royston, J.P. (1992a). Approximating the Shapiro-Wilk W-Test for Non-Normality. *Statistics and Computing* **2**, 117–119.

Royston, J.P. (1992b). Estimation, Reference Ranges and Goodness of Fit for the Three-Parameter Log-Normal Distribution. *Statistics in Medicine* **11**, 897–912.

Royston, J.P. (1992c). A Pocket-Calculator Algorithm for the Shapiro-Francia Test of Non-Normality: An Application to Medicine. *Statistics in Medicine* **12**, 181–184.

Royston, P. (1993). A Toolkit for Testing for Non-Normality in Complete and Censored Samples. *The Statistician* **42**, 37–43.

Rubinstein, R.Y. (1981). *Simulation and the Monte Carlo Method.* John Wiley & Sons, New York, 278 pp.

Ruffle, B., D.E. Burmaster, P.D. Anderson, and H.D. Gordon. (1994). Lognormal Distributions for Fish Consumption by the General U.S. Population. *Risk Analysis* **14**(4), 395–404.

Ryan, T.P. (1989). *Statistical Methods for Quality Improvement.* John Wiley & Sons, New York, 446 pp.

Saltelli, A., and J. Marivoet. (1990). Non-Parametric Statistics in Sensitivity Analysis for Model Output: A Comparison of Selected Techniques. *Reliability Engineering and System Safety* **28**, 229–253.

Sara, Martin N. (1994). *Standard Handbook for Solid and Hazardous Waste Facility Assessment.* Lewis Publishers, Boca Raton, FL. Chapter 11.

Sarhan, A.E., and B.G. Greenberg. (1956). Estimation of Location and Scale Parameters by Order Statistics for Singly and Doubly Censored Samples, Part I, The Normal Distribution up to Samples of Size 10. *Annals of Mathematical Statistics* **27**, 427–457.

Saw, J.G. (1961a). Estimation of the Normal Population Parameters Given a Type I Censored Sample. *Biometrika* **48**, 367–377.

Saw, J.G. (1961b). The Bias of the Maximum Likelihood Estimators of Location and Scale Parameters Given a Type II Censored Normal Sample. *Biometrika* **48**, 448–451.

Schaeffer, D.J., H.W. Kerster, and K.G. Janardan. (1980). Grab Versus Composite Sampling: A Primer for Managers and Engineers. *Journal of Environmental Management* **4**, 157–163.

Scheffé, H. (1959). *The Analysis of Variance.* John Wiley & Sons, New York, 477 pp.

Scheuer, E.M., and D.S. Stoller. (1962). On the Generation of Normal Random Vectors. *Technometrics* **4**(2), 278–281.

Schmee, J., D.Gladstein, and W. Nelson. (1985). Confidence Limits for Parameters of a Normal Distribution from Singly Censored Samples, Using Maximum Likelihood. *Technometrics* **27**(2), 119–128.

Schmoyer, R.L., J.J. Beauchamp, C.C. Brandt, and F.O. Hoffman, Jr. (1996). Difficulties with the Lognormal Model in Mean Estimation and Testing. *Environmental and Ecological Statistics* **3**, 81–97.

Schneider, H. (1986). *Truncated and Censored Samples from Normal Populations*. Marcel Dekker, New York, 273 pp.

Schneider, H., and L. Weisfeld. (1986). Inference Based on Type II Censored Samples. *Biometrics* **42**, 531–536.

Seiler, F.A. (1987). Error Propagation for Large Errors. *Risk Analysis* **7**(4), 509–518.

Seiler, F.A., and J.L. Alvarez. (1996). On the Selection of Distributions for Stochastic Variables. *Risk Analysis* **16**(1), 5–18.

Sengupta, S., and N. K. Neerchal. (1994). Classification of Grab Samples Using Composite Sampling: An Improved Strategy. *Calcutta Statistical Association Bulletin* **44**, 195–208.

Serfling, R.J. (1980). *Approximation Theorems of Mathematical Statistics*. John Wiley & Sons, New York, 371 pp.

Shaffner, G. (1999). *The Arithmetic of Life*. Ballantine Books, New York, 208 pp.

Shapiro, S.S., and C.W. Brian. (1981). A Review of Distributional Testing Procedures and Development of a Censored Sample Distributional Test. In Taillie, C., G.P. Patil, and B.A. Baldessari, eds. *Statistical Distributions in Scientific Work, Volume 5—Inferential Problems and Properties*, 1–24.

Shapiro, S.S., and R.S. Francia. (1972). An Approximate Analysis of Variance Test for Normality. *Journal of the American Statistical Association* **67**(337), 215–219.

Shapiro, S.S., and M.B. Wilk. (1965). An Analysis of Variance Test for Normality (Complete Samples). *Biometrika* **52**, 591–611.

Shea, B.L., and A.J. Scallon. (1988). Remark AS R72. A Remark on Algorithm AS 128. Approximating the Covariance Matrix of Normal Order Statistics. *Applied Statistics* **37**, 151–155.

Sheskin, D.J. (1997). *Handbook of Parametric and Nonparametric Statistical Procedures*. CRC Press, Boca Raton, FL, 719 pp.

Shlyakhter, A.I. (1994). An Improved Framework for Uncertainty Analysis: Accounting for Unsuspected Errors. *Risk Analysis* **14**(4), 441–447.

Shumway, R.H., A.S. Azari, and P. Johnson. (1989). Estimating Mean Concentrations under Transformations for Environmental Data with Detection Limits. *Technometrics* **31**(3), 347–356.

Silverman, B. W. (1986). *Density Estimation for Statistics and Data Analysis.* Chapman & Hall, New York.

Singh, A. (1993). Multivariate Decision and Detection Limits. *Analytica Chimica Acta* **277**, 205–214.

Singh, A.K., A. Singh, and M. Engelhardt. (1997). *The Lognormal Distribution in Environmental Applications.* EPA/600/R-97/006. Technology Support Center for Monitoring and Site Characterization, Technology Innovation Office, Office of Research and Development, Office of Solid Waste and Emergency Response, U.S. Environmental Protection Agency, Washington, D.C., December, 1997.

Size, W.B., ed. (1987). *Use and Abuse of Statistical Methods in the Earth Sciences.* Oxford University Press, New York, 169 pp.

Slob, W. (1994). Uncertainty Analysis in Multiplicative Models. *Risk Analysis* **14**(4), 571–576.

Smirnov, N.V. (1939). Estimate of Deviation between Empirical Distribution Functions in Two Independent Samples. *Bulletin Moscow University* **2**(2), 3–16.

Smirnov, N.V. (1948). Table for Estimating the Goodness of Fit of Empirical Distributions. *Annals of Mathematical Statistics* **19**, 279–281.

Smith, A.E., P.B. Ryan, and J.S. Evans. (1992). The Effect of Neglecting Correlations When Propagating Uncertainty and Estimating the Population Distribution of Risk. *Risk Analysis* **12**(4), 467–474.

Smith, E.P. (1994). Biological Monitoring: Statistical Issues and Models. In Patil, G.P., and C.R. Rao, eds., *Handbook of Statistics, Vol. 12: Environmental Statistics.* North-Holland, Amsterdam, Chapter 8, 243–261.

Smith, E.P., and K. Rose. (1991). Trend Detection in the Presence of Covariates: Stagewise Versus Multiple Regression. *Environmetrics* **2**(2), 153–168.

Smith, R.L. (1985). Maximum Likelihood Estimation in a Class of Nonregular Cases. *Biometrika* **72**(1), 67–90.

Smith, R.L. (1994). Use of Monte Carlo Simulation for Human Exposure Assessment at a Superfund Site. *Risk Analysis* **14**(4), 433–439.

Smith, R.M., and L.J. Bain. (1976). Correlation Type Goodness-of-Fit Statistics with Censored Sampling. *Communications in Statistics A—Theory and Methods* **5**(2), 119–132.

Snedecor, G.W., and W.G. Cochran. (1989). *Statistical Methods.* Eighth Edition. Iowa State University Press, Ames, IA, 503 pp.

Sokal, R.F., and F.J. Rolfe. (1981). *Biometry: The Principals and Practice of Statistics in Biological Research.* Second Edition. W.H. Freeman and Company, San Francisco, 859 pp.

Spiegelman, C.H. (1997). A Discussion of Issues Raised by Lloyd Currie and a Cross Disciplinary View of Detection Limits and Estimating Parameters That Are Often at or near Zero. *Chemometrics and Intelligent Laboratory Systems* **37**, 183–188.

Springer, M.D. (1979). *The Algebra of Random Variables.* John Wiley & Sons, New York, 470 pp.

Starks, T.H. (1988). *Evaluation of Control Chart Methodologies for RCRA Waste Sites.* Draft Report by Environmental Research Center, University of Nevada, Las Vegas, for Exposure Assessment Research Division, Environmental Monitoring Systems Laboratory—Las Vegas, Nevada. EPA Technical Report CR814342-01-3.

Stedinger, J. (1983). Confidence Intervals for Design Events. *Journal of Hydraulic Engineering* **109**(1), 13–27.

Stedinger, J.R. (1980). Fitting Lognormal Distributions to Hydrologic Data. *Water Resources Research* **16**(3), 481–490.

Stedinger, J.R., R.M. Vogel, and E. Foufoula-Georgiou. (1993). Frequency Analysis of Extreme Events. In Maidment, D.R., ed. *Handbook of Hydrology.* McGraw-Hill, New York, Chapter 18.

Stein, M. (1987). Large Sample Properties of Simulations Using Latin Hypercube Sampling. *Technometrics* **29**(2), 143–151.

Stephens, M.A. (1970). Use of the Kolmogorov-Smirnov, Cramér-von Mises and Related Statistics Without Extensive Tables. *Journal of the Royal Statistical Society, Series B* **32**, 115–122.

Stephens, M.A. (1986a). Tests Based on EDF Statistics. In D'Agostino, R. B., and M.A. Stevens, eds. *Goodness-of-Fit Techniques.* Marcel Dekker, New York, Chapter 4, pp. 97–193.

Stephens, M.A. (1986b). Tests Based on Regression and Correlation. In D'Agostino, R. B., and M.A. Stevens, eds. *Goodness-of-Fit Techniques.* Marcel Dekker, New York, Chapter 5, pp. 195–233.

Stokes, S.L. (1980). Estimation of Variance Using Judgment Ordered Ranked Set Samples. *Biometrics* **36**, 35–42.

Stolarski, R.S., A.J. Krueger, M.R. Schoeberl, R.D. McPeters, P.A. Newman, and J.C. Alpert. (1986). Nimbus 7 Satellite Measurements of the Springtime Antarctic Ozone Decrease. *Nature* **322**, 808–811.

Stoline, M.R. (1991). An Examination of the Lognormal and Box and Cox Family of Transformations in Fitting Environmental Data. *Environmetrics* **2**(1), 85–106.

Stoline, M.R. (1993). Comparison of Two Medians Using a Two-Sample Lognormal Model in Environmental Contexts. *Environmetrics* **4**(3), 323–339.

Suter, G.W. (1993). *Ecological Risk Assessment.* Lewis Publishers, Boca Raton, FL, 538 pp.

Taylor, J.K. (1987). *Quality Assurance of Chemical Measures.* Lewis Publishers, CRC Press, Boca Raton, FL, 328 pp.

Taylor, J.K. (1990). *Statistical Techniques for Data Analysis.* Lewis Publishers, Boca Raton, FL, 200 pp.

Thompson, K.M., and D.E. Burmaster. (1991). Parametric Distributions for Soil Ingestion by Children. *Risk Analysis* **11**(2), 339–342.

Thompson, K.M., D.E. Burmaster, and E.A.C. Crouch. (1992). Monte Carlo Techniques for Quantitative Uncertainty Analysis in Public Health Risk Assessments. *Risk Analysis* **12**(1), 53–63.

Tiago de Oliveira, J. (1963). Decision Results for the Parameters of the Extreme Value (Gumbel) Distribution Based on the Mean and Standard Deviation. *Trabajos de Estadistica* **14**, 61–81.

Tiago de Oliveira, J. (1983). Gumbel Distribution. In Kotz, S., and N. Johnson, eds. *Encyclopedia of Statistical Sciences.* John Wiley & Sons, New York. Volume 3, pp. 552–558.

Travis, C.C., and M.L. Land. (1990). Estimating the Mean of Data Sets with Nondetectable Values. *Environmental Science and Technology* **24**, 961–962.

Tsui, K.L., N.P. Jewell, and C.F.J. Wu. (1988). A Nonparametric Approach to the Truncated Regression Problem. *Journal of the American Statistical Association* **83**(403), 785–792.

USEPA. (1983). Standard for Remedial Actions at Inactive Uranium Processing Sites; Final Rule (40 CFR Part 192). *Federal Register* **48**(3), 590–604.

USEPA. (1987a). *Data Quality Objectives for Remedial Response Activities, Development Process.* EPA/540/G-87/003. U.S. Environmental Protection Agency, Washington, D.C.

USEPA. (1987b). *Data Quality Objectives for Remedial Response Activities, Example Scenario RI/FS Activities at a Site with Contaminated Soils and Ground Water.* EPA/540/G-87/004. U.S. Environmental Protection Agency, Washington, D.C.

USEPA. (1987c). List (Phase 1) of Hazardous Constituents for Ground-Water Monitoring; Final Rule. *Federal Register* **52**(131), 25942–25953 (July 9, 1987).

USEPA. (1988). Statistical Methods for Evaluating Ground-Water Monitoring from Hazardous Waste Facilities: Final Rule. *Federal Register* **53**, 39, 720–731.

USEPA. (1989a). *Methods for Evaluating the Attainment of Cleanup Standards, Volume 1: Soils and Solid Media.* EPA/230-02-89-042. Office of Policy, Planning, and Evaluation, U.S. Environmental Protection Agency, Washington, D.C.

USEPA. (1989b). *Statistical Analysis of Ground-Water Monitoring Data at RCRA Facilities, Interim Final Guidance.* EPA/530-SW-89-026. Office of Solid Waste, U.S. Environmental Protection Agency, Washington, D.C.

USEPA. (1989c). *Risk Assessment Guidance for Superfund, Volume 1: Human Health Evaluation Manual (Part A), Interim Final Guidance.* EPA/540/1-89/002. Office of Emergency and Remedial Response, U.S. Environmental Protection Agency, Washington, D.C.

USEPA. (1990). *Test Methods for Evaluating Solid Waste, 3rd Edition, Proposed Update I.* Office of Solid Waste and Emergency Response, U.S. Environmental Protection Agency, Washington, D.C.

USEPA. (1991a). *Data Quality Objectives Process for Planning Environmental Data Collection Activities, Draft.* Quality Assurance Management Staff, Office of Research and Development, U.S. Environmental Protection Agency, Washington, D.C.

USEPA. (1991b). *Risk Assessment Guidance for Superfund, Volume 1: Human Health Evaluation Manual, Supplemental Guidance/Standard Default Exposure Factors, Interim Final.* OSWER Directive 9285.6-03. Office of Emergency and Remedial Response, U.S. Environmental Protection Agency, Washington, D.C.

USEPA. (1991c). Solid Waste Disposal Facility Criteria: Final Rule. *Federal Register* **56**, 50978–51119.

USEPA. (1992a). *Characterizing Heterogeneous Wastes: Methods and Recommendations.* EPA/600/R-92/033. U.S. Environmental Protection Agency, Washington, D.C.

USEPA. (1992b). *Methods for Evaluating the Attainment of Cleanup Standards, Volume 2: Groundwater.* EPA/230-R-92-014. Office of Policy, Planning, and Evaluation, U.S. Environmental Protection Agency, Washington, D.C.

USEPA. (1992c). *Statistical Analysis of Ground-Water Monitoring Data at RCRA Facilities: Addendum to Interim Final Guidance.* Office of Solid Waste, U.S. Environmental Protection Agency, Washington, D.C. Currently available as part of: Statistical Training Course for Ground-Water Monitoring Data Analysis, EPA/530-R-93-003, which may be obtained through the RCRA Docket (202/260-9327).

USEPA. (1992d). *Supplemental Guidance to RAGS: Calculating the Concentration Term.* Publication 9285.7-081, May 1992. Intermittent Bulletin, Volume 1, Number 1. Office of Emergency and Remedial Response, Hazardous Site Evaluation Division, OS-230. Office of Solid Waste and Emergency Response, U.S. Environmental Protection Agency, Washington, D.C.

USEPA. (1994a). *Guidance for the Data Quality Objectives Process, EPA QA/G-4.* EPA/600/R-96/005. Office of Research and Development, U.S. Environmental Protection Agency, Washington, D.C.

USEPA. (1994b). *Statistical Methods for Evaluating the Attainment of Cleanup Standards, Volume 3: Reference-Based Standards for Soils and Solid Media.* EPA/230-R-94-004. Office of Policy, Planning, and Evaluation, U.S. Environmental Protection Agency, Washington, D.C.

USEPA. (1994c). *Use of Monte Carlo Simulation in Risk Assessments.* EPA/903-F-94-001. Hazardous Waste Management Division, U.S. Environmental Protection Agency, Region III, Philadelphia, PA.

USEPA. (1995a). *EPA Observational Economy Series, Volume 1: Composite Sampling.* EPA/230-R-95-005. Office of Policy, Planning, and Evaluation, U.S. Environmental Protection Agency, Washington, D.C.

USEPA. (1995b). *EPA Observational Economy Series, Volume 2: Ranked Set Sampling.* EPA/230-R-95-006. Office of Policy, Planning, and Evaluation, U.S. Environmental Protection Agency, Washington, D.C.

USEPA. (1995c). *EPA Risk Characterization Policy and Guidance.* Memorandum from Carol M. Browner, Administrator, U.S. Environmental Protection Agency, March 21, 1995.

USEPA. (1996a). *Guidance for Data Quality Assessment: Practical Methods for Data Analysis, EPA QA/G-9, QA96 Version.* EPA/600/R-96/084, July 1996. Office of Research and Development, U.S. Environmental Protection Agency, Washington, D.C.

USEPA. (1996b). *Soil Screening Guidance: User's Guide.* EPA/540/R-96/018, PB96963505. Office of Emergency and Remedial Response, U.S. Environmental Protection Agency, Washington, D.C., April, 1996.

USEPA. (1996c). *Soil Screening Guidance: Technical Background Document.* EPA/540/R-95/128, PB96963502. Office of Emergency and Remedial Response, U.S. Environmental Protection Agency, Washington, D.C., May, 1996.

USEPA. (1997a). *Guiding Principles for Monte Carlo Analysis*. EPA/630/R-97/001. Risk Assessment Forum, U.S. Environmental Protection Agency, Washington, D.C.

USEPA. (1997b). *Policy for Use of Monte Carlo Analysis in Risk Assessment, With Attachment (Draft)*. Memorandum from William P. Wood, January 29, 1997. U.S. Environmental Protection Agency, Washington, D.C.

USEPA. (1998a). *Guidance for Data Quality Assessment: Practical Methods for Data Analysis, EPA QA/G-9, QA97 Version*. EPA/600/R-96/084, January 1998. Office of Research and Development, U.S. Environmental Protection Agency, Washington, D.C.

USEPA. (1998b). *Guidance for Quality Assurance Project Plans, EPA QA/G-5*. EPA/600/R-98/018. Office of Research and Development, U.S. Environmental Protection Agency, Washington, D.C.

van Belle, G., and J.P. Hughes. (1984). Nonparametric Tests for Trend in Water Quality. *Water Resources Research* **20**(1), 127–136.

van Belle, G., and D.C. Martin. (1993). Sample Size as a Function of Coefficient of Variation and Ratio of Means. *The American Statistician* **47**(3), 165–167.

Venables, W.N., and B.D. Ripley. (1999). *Modern Applied Statistics with S-PLUS*. Third Edition. Springer-Verlag, New York, 501 pp.

Verrill, S., and R.A. Johnson. (1987). The Asymptotic Equivalence of Some Modified Shapiro-Wilk Statistics—Complete and Censored Sample Cases. *The Annals of Statistics* **15**(1), 413–419.

Verrill, S., and R.A. Johnson. (1988). Tables and Large-Sample Distribution Theory for Censored-Data Correlation Statistics for Testing Normality. *Journal of the American Statistical Association* **83**, 1192–1197.

Vogel, R.M. (1986). The Probability Plot Correlation Coefficient Test for the Normal, Lognormal, and Gumbel Distributional Hypotheses. *Water Resources Research* **22**(4), 587–590. (Correction, *Water Resources Research* **23**(10), 2013, 1987.)

Vogel, R.M., and N.M. Fennessey. (1993). L Moment Diagrams Should Replace Product Moment Diagrams. *Water Resources Research* **29**(6), 1745–1752.

Vogel, R.M., and D.E. McMartin. (1991). Probability Plot Goodness-of-Fit and Skewness Estimation Procedures for the Pearson Type 3 Distribution. *Water Resources Research* **27**(12), 3149–3158.

Vose, D. (1996). Quantitative Risk Analysis: *A Guide to Monte Carlo Simulation Modelling*. John Wiley & Sons, New York, 328 pp.

Wald, A., and J. Wolfowitz. (1946). Tolerance Limits for a Normal Distribution. *Annals of Mathematical Statistics* **17**, 208–215.

Walsh, J. (1996). *True Odds: How Risk Affects Your Everyday Life*. Merritt Publishing, Santa Monica, CA, 401 pp.

Webster, R., and M.A. Oliver. (1990). *Statistical Methods in Soil and Land Resource Survey*. Oxford University Press, New York, 316 pp.

Weinstein, N.D., P.M. Sandman, and W.K. Hallman. (1994). Testing a Visual Display to Explain Small Probabilities. *Risk Analysis* **14**(6), 895–896.

Weisberg, S. (1985). *Applied Linear Regression*. Second Edition. John Wiley & Sons, New York, 324 pp.

Weisberg, S., and C. Bingham. (1975). An Approximate Analysis of Variance Test for Non-Normality Suitable for Machine Calculation. *Technometrics* **17**(1), 133–134.

Wicksell, S.D. (1917). On Logarithmic Correlation with an Application to the Distribution of Ages at First Marriage. *Medd. Lunds. Astr. Obs.* **84**, 1–21.

Wilk, M.B., and R. Gnanadesikan. (1968). Probability Plotting Methods for the Analysis of Data. *Biometrika* **55**, 1–17.

Wilk, M.B., and S.S. Shapiro. (1968). The Joint Assessment of Normality of Several Independent Samples. *Technometrics* **10**(4), 825–839.

Wilks, S.S. (1941). Determination of Sample Sizes for Setting Tolerance Limits. *Annals of Mathematical Statistics* **12**, 91–96.

WSDOE. (1992). *Statistical Guidance for Ecology Site Managers*. Washington State Department of Ecology, Olympia, WA.

WSDOE. (1993). *Statistical Guidance for Ecology Site Managers, Supplement S-6: Analyzing Site or Background Data with Below-Detection Limit or Below-PQL Values (Censored Data Sets)*. Washington State Department of Ecology, Olympia, WA.

Zacks, S. (1970). Uniformly Most Accurate Upper Tolerance Limits for Monotone Likelihood Ratio Families of Discrete Distributions. *Journal of the American Statistical Association* **65**, 307–316.

Zar, J.H. (1999). *Biostatistical Analysis*. Fourth Edition. Prentice-Hall, Upper Saddle River, NJ.

Zorn, M.E., R.D. Gibbons, and W.C. Sonzogni. (1997). Weighted Least-Squares Approach to Calculating Limits of Detection and Quantification by Modeling Variability as a Function of Concentration. *Analytical Chemistry* **69**, 3069–3075.

Index

E

F

G

H

I

J